Youngsters Solving Mathematical Problems with Technology

The Results and Implications of the Problem@Web Project

MATHEMATICS EDUCATION IN THE DIGITAL ERA
Volume 5

Series Editors:
Dragana Martinovic, University of Windsor, ON, Canada
Viktor Freiman, Université de Moncton, NB, Canada

Editorial Board:
Marcelo Borba, State University of São Paulo, São Paulo, Brazil
Rosa Maria Bottino, CNR – Istituto Tecnologie Didattiche, Genova, Italy
Paul Drijvers, Utrecht University, Utrecht, the Netherlands
Celia Hoyles, University of London, London, UK
Zekeriya Karadag, Giresun Üniversitesi, Giresun, Turkey
Stephen Lerman, London South Bank University, London, UK
Richard Lesh, Indiana University, Bloomington, USA
Allen Leung, Hong Kong Baptist University, Hong Kong
John Mason, Open University, UK
Sergey Pozdnyakov, Saint-Petersburg State Electro Technical University, Saint-Petersburg, Russia
Ornella Robutti, Università di Torino, Torino, Italy
Anna Sfard, Michigan State University, USA & University of Haifa, Haifa, Israel
Bharath Sriraman, University of Montana, Missoula, USA
Anne Watson, University of Oxford, Oxford, UK

More information about this series at http://www.springer.com/series/10170

Susana Carreira • Keith Jones • Nélia Amado
Hélia Jacinto • Sandra Nobre

Youngsters Solving Mathematical Problems with Technology

The Results and Implications of the Problem@Web Project

Susana Carreira
Universidade do Algarve
Faro, Portugal

Keith Jones
University of Southampton
Southampton, UK

Nélia Amado
Universidade do Algarve
Faro, Portugal

Hélia Jacinto
UIDEF, Instituto de Educação
Universidade de Lisboa
Lisboa, Portugal

Sandra Nobre
UIDEF, Instituto de Educação
Universidade de Lisboa
Lisboa, Portugal

ISSN 2211-8136 ISSN 2211-8144 (electronic)
Mathematics Education in the Digital Era
ISBN 978-3-319-24908-7 ISBN 978-3-319-24910-0 (eBook)
DOI 10.1007/978-3-319-24910-0

Library of Congress Control Number: 2015956101

Springer Cham Heidelberg New York Dordrecht London
© Springer International Publishing Switzerland 2016
This work is subject to copyright. All rights are reserved by the Publisher, whether the whole or part of the material is concerned, specifically the rights of translation, reprinting, reuse of illustrations, recitation, broadcasting, reproduction on microfilms or in any other physical way, and transmission or information storage and retrieval, electronic adaptation, computer software, or by similar or dissimilar methodology now known or hereafter developed.
The use of general descriptive names, registered names, trademarks, service marks, etc. in this publication does not imply, even in the absence of a specific statement, that such names are exempt from the relevant protective laws and regulations and therefore free for general use.
The publisher, the authors and the editors are safe to assume that the advice and information in this book are believed to be true and accurate at the date of publication. Neither the publisher nor the authors or the editors give a warranty, express or implied, with respect to the material contained herein or for any errors or omissions that may have been made.

Printed on acid-free paper

Springer International Publishing AG Switzerland is part of Springer Science+Business Media (www.springer.com)

Foreword

The Digital Revolution in Society and in School

I want to turn to a metaphor I used in a publication in 2007 in which I compared the information society to a kind of black hole that was involving all activities that, even just once or in a small part of their scopes, needed to use a technology that would activate bits.

I also argued that school should position itself as a driver of change and as fundamental for the development of the information and knowledge society. If the school is able to accept and promote within it the new available tools and methods, students who take advantage of them will surely become better prepared citizens.

Similarly, back in the beginning of the century, Manuel Castells has advocated that there is a world before and a world after the emergence of the Internet and that this was (is) a revolution comparable to the first and second industrial revolutions. Eventually it will be even a more vehement and striking one in the life and well-being of citizens.

The professions have changed in their environments, technologies, techniques and methods. In fact, there are jobs that have changed more than others. Interestingly, the teaching profession was one of the least changed. Today, the classroom differs little from the classroom of the beginning of the twentieth century. There are changes actually, but those changes have not been very apparent. Therefore, it is vital and imperative that changes not only happen but become increasingly evident.

Many studies that we may find and confront on the added value of digital technologies in the learning processes already give us a very favourable image of its use. This is not only because students learn the content better when undergoing new paradigms of learning, but mainly because they acquire skills pertaining to a digital literacy that is clearly needed in a competitive world as ours is and will be.

These research studies seek to understand whether or not there is a positive impact of ICT on student learning. A central question is whether, by using ICT, students learn more and better. Existing literature is not very conclusive about the effectiveness of the use of digital technologies, but it confirms that there is greater involvement and motivation of the actors, especially the students. Even parents, when requested to express their opinion, indicate that children have a more positive attitude towards homework and begin to talk more about school activities when digital technologies are involved.

Schools cannot stop the digital world from entering their domain or the classroom. The students' portable technologies (mobile phones and tablets) may not be forbidden in the name of the school and of academic order that is more or less self-centred and very little open to what is happening outside. Students cannot be denied what is good about the information society to which they have access beyond the school gates.

Policies need to change, though the effort has to be huge: global incentive policies and local policies facilitating and promoting innovation and creating learning schools will be the steps of a journey that we are already undertaking. However, there should be an eye to the fact that technologies do not have just a positive face and that its use is not deprived of risk. Now, it is mainly up to the school to alert the children and adolescents about the risks that exist when using the Internet and its various services. Virus attacks, access to unreliable information and contacts with unidentified persons are some of the dangers. Furthermore, we must raise awareness that in this new Internet world full of information, we should use it according to the principles of ethics and responsibility, by realising that any content has its author and therefore has to be properly referenced when used on schoolwork, whatever may be its breadth.

Definitely, the school cannot remain outside of the reality, pretending that the world out there does not change.

The Cyber Youngsters

In the digital world, we are concerned with global and local policies, widespread access, digital divides, trained actors, etc. Interestingly, when we speak of actors, there are very few times that we identify them with students. Since Prensky coined the term digital natives, it has been assumed that youngsters were equipped with skills provided by these instruments of the information society. They type at impressive speeds on their mobile phones, they talk on Skype with friends, and they write on their Facebook walls and on their friends' whom they have never seen. As Tapscott would say, these youngsters are the Net Generation.

Our students are "digital natives" because they were born in a digital age, but what is happening is that, like the "digital immigrants", they do not truly dominate much of the affordances of the tools made available by the information society and the Internet. This raises the issue of what we should do so that students reach the "digital wisdom", as Prensky later rethought the previous terminology.

Today's educational paradigms may not be the same as they were a few decades ago. Nobody dares to say that the so-called traditional methods have no place. But in fact, the classroom should be ever more a place where students feel comfortable, with a real desire to learn and with access to tools that provide them with knowledge. ICT have to actually enter the school and, further than the school, the student's learning process whether inside or outside the classroom environment.

When we look carefully, as has been shown in some research we carried out with students in the first years of higher education, their digital skills are much lower than we would suppose. In essence they have communication skills in social networks and research skills through the use of Google search engines. However, if students are asked to perform proper and rigorous text editing, or calculations in Excel, or even the production of a good PowerPoint presentation, the situation gets complicated. Indeed, the situation worsens still more when we ask them if they have, or they use, a blog, or a website, or a portfolio of their works in digital form, or if they know how to use conceptual maps or software for the production of timelines.

Basically, the conclusion we, as teachers, reach is that these students, who we might label as digital natives and who we presume would know more than we do about these new technologies, have not properly acquired their digital literacy after all. This is not their fault; it is more the fault of adults who have assigned labels to them and expected skills from them that in reality are not spontaneously learned. Even the multitasking features to which Don Tapscott refers should be considered carefully. Our digital natives may be doing many things simultaneously, but sometimes the degree of commitment may not be very deep. Contrarily, Tapscott refers in a somewhat candid and contextualised way to this new generation having positive preferences. According to him, the youth of the Net Generation cherish freedom and the freedom of their choices; they want to adapt things in terms of their ownership; they act collaboratively and prefer talking to listening to a teacher's lecture. They prefer to work and study with associated fun. This somewhat romantic version cannot be extrapolated beyond the context of observation where it was built. Many young people around the world do not use and do not feel technology this way—far from it, because they live in completely different contexts. This is why schools have the role of levelling between technology "haves" and "have-nots", between those who know and those who also want to know.

Projects such as the Problem@Web, which is embodied in this book, can be used to safely improve these skills not only of youngsters to whom the project is addressed, but also of their teachers and parents themselves, who will need to catch up and understand that there are now new ways of overcoming the challenges of learning mathematics or any other area of knowledge.

I am convinced that digital literacy and the proper use of digital tools can also be a family learning context, provided that parents have the attitude and the digital skills. If children see their parents working on computers, reading magazines and digital newspapers on their tablets and seeking answers to questions by searching with Google or on Wikipedia, then, somehow, they are being subjected to a model that will naturally condition and enhance their future behaviour.

The Immigrant Teachers, Innovation and Comfort Zone

Being a teacher in a digital world is becoming an arduous task. Students are full of technology in their personal worlds but attend classrooms that often, too often, offer few technological opportunities. Teachers in Portugal, for example, today show a high rate of computer ownership and daily use (over 95 %). Yet the movement of technology from the personal sphere to the learning environment becomes a complicated process and sometimes with less than positive results. Teachers themselves state their reasons, the first of which is the lack of training and the second the lack of specific software for their school subjects. In fact, it seems that these reasons do not have much underlying basis. One reason why this move does not occur has to do with what some have called a "moral panic" of teachers who feel that they know less than their students about the digital world. This causes many of them to have feelings of unwarranted insecurity.

Even so, many teachers make some effort to learn how to use the personal computer and know how to do the basics with productivity tools. They venture into the use of the PowerPoint as a support tool for their teaching. Yet they may not be putting digital tools to the service of the genuine construction of their students' learning. Teachers know a lot more technology than they themselves think. After all, they are proficient with Word, PowerPoint and Excel. And it is enough to engage students in activities involving at least these tools.

From the analysis of many of the mathematical problems by the Problem@Web project, it appears that students invent and reinvent the use of digital technology for expressing their mathematical thinking and solving the problems, often in different ways. This raises the question of why teachers are not more daring in these approaches. On the one hand, it can be said that there is some fear of failure and showing digital ignorance in front of the so-called digital natives. On the other hand, there is some difficulty in getting out of the comfort zone and being innovative, to innovate in practices and strategies not confined to teaching but aiming at creating contexts and learning pathways where technology can play at least a significant role.

This project posits, however, a fascinating idea. When there is freedom of choice of digital tools, students use those that they know better or they are more comfortable with, but even those who use the same software feature differentiated paths and solutions. Here the mathematical thinking is aided by the intelligent and innovative use of digital technologies. Simultaneously, the context pushes the students forward and enhances their skills, not only mathematical but also digital. The Problem@Web evidence shows that our students, very much involved in communication contexts and social networks, are also able to use other digital tools once they realise that this will help their reasoning and the presentation of their work.

And here is an important role for teachers—in daily practice to serve as a model of the personal use they make of ICT and constantly to challenge their students to conduct their work with the aid of digital tools, whenever appropriate.

As it says in this book, the spontaneous use of technology changes and reshapes the solving of mathematical problems. However, this spontaneity, which is found useful and fertile, can and should also be induced by the daily practice of teachers.

Are Mobile Technologies the Future?

There are interesting references in the book on the progress that students have made in the use of technology throughout the various editions of the online competitions that were studied. From scans of the resolutions made on paper, students went to utility programs. Word processors, drawing programs and spreadsheets had the honour of extensive use, especially from the older students. Other more specific software such as GeoGebra has also appeared. So what does the future hold?

Today we are aware that a new world opens to education with the rise of the first technologies capable of giving meaning to the concept of mobile learning. This is beyond what apparently the laptop failed to achieve—portability and the existence of a countless number of applications able to solve problems of various kinds and related to almost all content matters.

Tablets are enabling as never before the existence of one computer per student, allowing the teacher, for the first time, to take advantage of this tremendous educational potential. It does not matter the model of access to technology; what matters is that its use is transparent and oriented to solve problems more effectively and with increased quality.

This development, which is expected in the area of mobile technologies, will surely change the way the students of the future will present their solutions to the problems suggested in the Problem@Web project, being certain that such solutions will be innovative, creative and often unexpected.

Lisbon, Portugal
May 3, 2015

José Reis Lagarto

Preface

This book brings together the development of a theoretical argument and its translation into an extensive empirical context as a result of the research work carried out by the Problem@Web project. Through the 3 years of the project, we studied youngsters engaged in mathematical problem-solving using digital technologies of their choice and to which they had personal access. Problem@Web, the compact name given to our project, alludes directly to the research context, that of youngsters solving mathematical problems in their own time and with their own choice of digital technology.

The origins of the project arose in the experiences during the SUB12 and SUB14 web-based mathematical problem-solving competitions that have been taking place annually in the south of Portugal since 2005. Organised by the Mathematics Department of the Faculty of Sciences and Technology of the University of Algarve, SUB12 aims at 5th and 6th graders (10–12-year-olds), while SUB14 addresses students in 7th and 8th grade (12–14-year-olds). These are examples of the kind of mathematical competitions that it is important to know more about; they run online, they are inclusive aiming to support wide participation, and they involve what we call "moderate" mathematical challenges. We expand on all these features in the book.

This book is designed to offer new views that extend the knowledge on how today's youngsters tackle mathematics problems using the technology they have at their disposal. Through our analysis, we want to see how the current generation, who are growing up with digital technologies, have skills and performances that might be quite different from earlier generations. What we uncover is at odds with what some may think; the young people we studied are highly competent in solving mathematical problems through using digital technology and seem to have knowledge that is distinct to their parents and many older educated people.

Opening a window on the world of these young people through offering them challenges where the digital technology is the mediator turned out to be a way to reach them successfully. Mathematical challenges and beyond-school online competitions, such as SUB12 and SUB14, can succeed with young people. In doing so the data we gathered have provided an unparalleled opportunity to better know how youngsters engage in mathematical problem-solving and express their mathematical thinking. This book offers readers the opportunity to know the richness and quality of mathematical knowledge produced by young people with the digital technologies they freely choose to use.

Primarily looking at a beyond-school context where students explore their "natural" learning resources, some of which may not always be available at school, we investigate what young people show as their spontaneous ways to express ideas and mathematical thinking. We believe that this knowledge can make a valuable contribution to understanding, and foreseeing, the school and the learning of mathematics in a noticeably digital age.

In the first of eight chapters, we begin by providing an introduction to research relating to youngsters' mathematical problem-solving with technology, together with an overview of the Problem@Web project including its rationale, aims and methods. The chapter also explains the methodological procedures developed in several of the subsequent chapters in which we concentrate on addressing our analysis of real data from youngsters mathematical problem-solving.

The second chapter presents fieldwork data mainly from interviews and digital material on the experiences of the youngsters who were involved in problem-solving within the web-based mathematical competitions. It offers a portrait of such youngsters as technology users, namely, when they utilise commonly available technological tools; it also illustrates the ways in which their technological competences are placed at the service of their mathematical problem-solving and expressing.

The third chapter presents additional data from interviews on the perspectives of the teachers regarding their students' participation in beyond-school projects that are directly related to the use of technologies in mathematical problem-solving, namely, web-based competitions. Issues related to students' unforeseen and creative strategies of solving problems and communicating their solutions and, in general, to the development of new teachers' views on these youngsters in the mathematics classroom are addressed.

In Chap. 4 we develop a theoretical argument about the unity between solving and expressing in problem-solving as a central construct that can be meaningfully correlated with the inseparability between the subject and the digital tool. This theoretical stance evolved from concepts such as humans with media and coaction. In framing the study of a specific and relatively new phenomenon, that of youngsters solving mathematical problems with the digital technologies of their choice, the theoretical tools and constructs that are reviewed and discussed lead to creating new constructs that we use to guide a better interpretation of what young people are able to do in a digital communication context (in the case of the Problem@Web project, finding the solution to a mathematical problem and its explanation within the scope of an online mathematical competition). Our aim is to understand how the youngsters find effective ways to achieve the solution of a problem and to communicate it mathematically, based on the digital resources they have at their disposal in their daily life, most cases in their home environment but also in school, including in the mathematics classroom.

Chapter 5 develops around two fundamental ideas, namely, (1) that the perception of the affordances of a certain digital tool is essential to solving mathematical problems with that particular technology and (2) that the activity thus undertaken stimulates different mathematising processes which, in turn, result in different conceptual models. Looking thoroughly, from an interpretative perspective, at four solutions to a particular geometry problem completed by youngsters who decided to use the dynamic geometry software *GeoGebra* at some point in their solving activity, our main purpose is to illustrate the ways in which the same tool affords different approaches to the problem in terms of the conceptual models developed by the youngsters for studying and justifying the invariance of the area of a triangle. Their different ways of dealing with the tool and with mathematical knowledge are interpreted as instances of youngsters with media engaged in a solving-with-*GeoGebra* activity, enclosing a range of procedures brought forth by the symbioses between the affordances of *GeoGebra* and the youngsters' aptitudes. The evidence shows that different youngsters solving the same problem with the same media and recognising a relatively similar set of affordances of the tool produce different digital solutions, but they also generate qualitatively different conceptual models for, in this case, the invariance of the area of a geometric shape.

In Chap. 6 we describe and analyse a number of examples of 7th and 8th graders presenting diverse ways of expressing their mathematical thinking in solving algebraic problems with a spreadsheet. Our research purposes are concerned with youngsters' approaches to situations where quantity variation is involved in finding an unknown value under a set of conditions that frame a problem situation. The use of the spreadsheet is thoroughly examined with the aim of highlighting the nature of problem-solving and expressing in the digital tool context as compared to the formal algebraic context; moreover, the ways in which students take advantage of the tool (being guided by and also guiding the spreadsheet distinctive forms of organising and performing variation in columns and cells) are important indicators of their algebraic thinking within the problem-solving activity.

Chapter 7 focuses on a motion problem that concerns the co-variation of displacement and time in a relative motion situation. The decision to focus on a motion problem has to do with the fact that a problem involving motion—while relating the variables space, time and speed—requires some kind of understanding of the dynamic nature of the problem situation and finding suitable models for their representation. We therefore look at the problem-solving and expressing of the youngsters when facing a motion problem, especially how most participants use some form of digital medium to express their thinking.

The final chapter summarises the overall findings of the project and considers the implications. Here we reveal how the youngsters that we studied had domain over a set of general-use digital tools and while they were less aware of digital resources with a stronger association with mathematics, they were able to gain many capabilities by tackling the mathematical problems and seeking expeditious, appropriate and productive ways of expressing their mathematical thinking. In particular, we review how they were able to harness their technological skills while simultaneously developing and improving their capacity to create and use a range of mathematical representations.

In sum, we present in this book an in-depth study of youngsters' mathematical problem-solving strategies and approaches that they demonstrate as they tackle mathematical problems in their own time and with digital resources of their own choice. Overall, our book provides the following:

- Numerous examples of moderately challenging mathematical problems.
- Many instances of student solutions, together with the students' explanations of how they achieved their solution; these student solutions are both a revelation and a valuable resource showing what youngsters can do in their own time and with their own choice of technology.
- A well-developed theoretical framing that integrates the use of technology into mathematical problem-solving.
- Insightful analysis of the young participants and their teachers and families and of the youngsters' mathematical problem-solving; the latter involving the mathematics of invariance, variation and co-variation.

It is our hope that this book contributes to the continuous development of research in mathematical problem-solving by unveiling the actual ways in which young students engage with challenging mathematical problems with the digital technological devices of their choice.

It is timely to acknowledge that the main ideas in this book are the outcome of the joint work of the authors while members of the Problem@Web project. Particular contributions were developed by Hélia Jacinto and Sandra Nobre based on the research they developed within their doctoral dissertations at the Institute of Education of the University of Lisbon. The book could not be produced without the support of the Fundação para a Ciência e Tecnologia that funded the Problem@Web project under the grant Nr PTDC/CPE-CED/101635/2008 and of the University of Algarve and the Institute of Education of the University of Lisbon which hosted the Problem@Web project. We would like to express our special thanks to Isa Martins, Jaime Carvalho e Silva, Juan Rodriguez, Nuno Amaral, Rosa Tomás Ferreira and Sílvia Reis who were members of the project team for their contributions, commitment and friendship throughout the project. We wish to thank all the teachers, the parents and the participants in SUB12 and SUB14 mathematical competitions who always supported us and so significantly engaged in this project; plus we thank those who took photographs during the competition finals and agreed for some of these to appear in this book. We would also like to thank Melissa James, from Springer, for her invaluable help and enthusiasm in pursuing this book; Dragana Martinovic and Viktor Freiman, editors of the MEDE Series; and Gerry Stahl for his insightful comments when reviewing the chapters of the book. We are honoured that José Lagarto (Universidade Católica Portuguesa, Portugal) provides the foreword and that Mónica Villarreal (Universidad de Córdoba, Argentina) the afterword.

Susana Carreira
Keith Jones
Nélia Amado
Hélia Jacinto
Sandra Nobre

Contents

1 **Mathematical Problem-Solving with Technology:**
   An Overview of the Problem@Web Project 1
   1.1 Introduction 1
   1.2 Young People with Technology 2
   1.3 Young People's Mathematical Problem-Solving
        with Technology 4
   1.4 The Research Focus 6
   1.5 The SUB12 and SUB14 Mathematics Competitions 8
   1.6 Methodological Issues 15
   1.7 Concluding Comments 17
   References 18

2 **Youngsters Solving Mathematical Problems with Technology:**
   Their Experiences and Productions 21
   2.1 Introduction 21
   2.2 The Participants in the Mathematical Competitions SUB12
        and SUB14 22
   2.3 The Participants and the Use of Digital Technologies 26
   2.4 The Participants' Productions with Digital Technologies 29
        2.4.1 From the Use of Paper and Pencil to Writing
               with Word and Excel 29
        2.4.2 The Use of Tables 35
        2.4.3 The Use of Images and Diagrams 39
        2.4.4 The Use of Numerical Software 44
        2.4.5 The Use of Geometrical Software 45
   2.5 Concluding Comments 51
   References 53

3 Perspectives of Teachers on Youngsters Solving Mathematical Problems with Technology 55
- 3.1 Introduction 55
- 3.2 The Role of the Teachers in the Mathematical Competitions 57
  - 3.2.1 The Support of the Teachers: From the First Round to the Final 59
  - 3.2.2 The Social Part of the Competitions: The Meeting at the Final 62
- 3.3 Perspectives of Teachers About the Mathematical Competitions SUB 12 and SUB14 63
- 3.4 Mathematical Communication: An Additional Challenge 72
- 3.5 The Use of Technology: The Sharing of Experiences Between Teachers and Students 74
- 3.6 Overview and Conclusion 79
- References 80

4 Theoretical Perspectives on Youngsters Solving Mathematical Problems with Technology 83
- 4.1 The Theoretical Stance 83
- 4.2 Problem-Solving as Mathematisation 85
- 4.3 Problem-Solving as Expressing Thinking 88
  - 4.3.1 Expository Discourse in Problem-Solving 90
  - 4.3.2 Technology Used for Expressing Thinking in Problem-Solving 93
- 4.4 Multiple External Representations 96
- 4.5 Humans-with-Media and Co-action with Digital Tools 98
- 4.6 An Outlook 106
- References 108

5 Digitally Expressing Conceptual Models of Geometrical Invariance 113
- 5.1 Main Theoretical Ideas 113
  - 5.1.1 Perceiving Affordances of Digital Tools 114
  - 5.1.2 The Indivisibility Between the Subject and the Context 117
  - 5.1.3 Humans-with-Media Mathematising 119
  - 5.1.4 Mathematisation with Dynamic Geometry Software 120
- 5.2 Context and Method 122
- 5.3 Data Analysis 124
  - 5.3.1 The Problem: Building a Flowerbed 124
  - 5.3.2 Zooming in: The Participants' Productions 125
  - 5.3.3 Zooming Out: Comparing and Contrasting 135
- 5.4 Discussion and Conclusion 138
- References 139

6	**Digitally Expressing Algebraic Thinking in Quantity Variation**		141
	6.1	Main Theoretical Ideas	141
		6.1.1 Digital Representations in the Spreadsheet	142
		6.1.2 Algebraic Thinking	144
		6.1.3 Problem-Solving with the Spreadsheet and the Development of Algebraic Thinking	145
		6.1.4 Expressing Algebraic Thinking and Co-action with the Spreadsheet	146
	6.2	Context and Method	147
	6.3	Data Analysis	149
		6.3.1 The First Problem: The Treasure of King Edgar	149
		6.3.2 The Second Problem: The Opening of the Restaurant "Sombrero Style"	160
	6.4	Discussion and Conclusion	168
	References		171
7	**Digitally Expressing Co-variation in a Motion Problem**		173
	7.1	Main Theoretical Ideas	173
		7.1.1 Co-variation and Modelling Motion	175
		7.1.2 Visualisation in Motion Problems	178
	7.2	Context and Method	180
	7.3	Data Analysis	182
		7.3.1 The Experts' Solutions to the Problem	182
		7.3.2 Definition of Categories	186
	7.4	Analysis of the Students' Solutions to the Problem	187
		7.4.1 Conceptual Models Involved in the Participants' Problem-Solving and Expressing	188
		7.4.2 Forms of Representation in Students' Digital Productions	191
	7.5	Discussion and Conclusion	204
	References		207
8	**Youngsters Solving Mathematical Problems with Technology: Summary and Implications**		209
	8.1	Introduction	209
	8.2	The Problem@Web Project	210
	8.3	The Youngsters Solving Mathematical Problems with Technology	214
	8.4	The Perspectives of the Youngsters' Teachers	218
	8.5	Theoretical Framework	223
	8.6	Digitally Expressing Mathematical Problem-Solving	225
		8.6.1 Digitally Expressing Conceptual Models of Geometrical Invariance	226

	8.6.2	Digitally Expressing Algebraic Thinking in Quantity Variation	227

        8.6.2 Digitally Expressing Algebraic Thinking in Quantity Variation 227
        8.6.3 Digitally Expressing Co-variation in a Motion Problem 230
  8.7 Discussion of the Findings 232
  8.8 Implications and Suggestions for Further Research 235
  References 237

Afterword 241

About the Authors 245

Index 249

Chapter 1
Mathematical Problem-Solving with Technology: An Overview of the Problem@Web Project

Abstract Today's youngsters are growing up in an era of rapidly advancing digital technologies. While young people in this generation are undoubtedly active users of digital technologies, the issue of whether their digital competency levels are necessarily well developed is a topic of debate. This chapter provides an introduction to, and an overview of, the Problem@Web project, a project that grew out of our interest in understanding how Portuguese youngsters participated in two online mathematical problem-solving competitions. These online competitions have allowed youngsters in any suitable place, and at any suitable time, to engage themselves in tackling mathematical problems by utilising solving strategies with any digital tools that they have available. During the project, we analysed numerous problem solutions submitted throughout three editions of the competitions and interviewed a sample of young participants, mathematics teachers and youngsters' parents and relatives. The chapter captures the contribution that the Problem@Web project makes to understand youngsters' mathematical problem-solving with technology.

Keywords Digital technologies • Mathematical problem-solving • Online mathematics competitions • Problem@Web project

1.1 Introduction

We live in an era of deep changes mediated by digital technologies. In recent times, there has been the launch of the smartphone and the tablet computer. Not so long ago, these words would have been written using a typewriter—until the advent of digital word processing which little by little began to enter into working lives and routines and then into the home. Young people born more recently have not felt these changes in the same way as their parents; today's youngsters were born, and are growing up, in the midst of a rapidly advancing technological environment. Today's youngsters did not witness the emergence of digital technologies; they were born in the midst of them and with them. Digital technologies are part of life for many of the generation that has variously been called digital natives (Prensky, 2001,

2006), Generation M (Roberts & Foehr, 2005) or the Net Generation (Tapscott, 1998, 2009). What these terms imply is that young people across many parts of the world are among the most active users of digital technologies.

Such an implication is confirmed by evidence from, for example, the 2013 *International Computer and Information Literacy Study* (ICILS) (Fraillon, Ainley, Schulz, Friedman, & Gebhardt, 2014) which found that today's youngsters make "widespread and frequent use of digital technologies when outside school" (p. 251). Grade 8 students (aged 13–14) in most of the ICILS participating countries typically said that they had been using computers for 5 years or more and reported using digital technologies for "study, communication, information exchange, and recreation" (ibid, p. 164). Such findings accord with Tapscott's (2009) claim that, at any one time, today's youngsters may be doing as many as five things simultaneously: texting friends, downloading music, uploading videos, watching a movie on a small screen and doing who-knows-what on social media. Such evidence might suggest that today's youngsters are active processors of information, accomplished computer gaming strategists and proficient social communicators.

Yet while young people are undoubtedly active users of digital technologies, the issue of whether their competency levels are necessarily well developed is being queried (see, e.g. Bennett, Maton, & Kervin, 2008; Helsper & Eynon, 2010). As Bennett et al., (2008, pp. 783–784) conclude, "The picture beginning to emerge from research on young people's relationships with technology is much more complex than the digital native characterisation suggests" such that what is needed is "considered and rigorous investigation that includes the perspectives of young people and their teachers, and genuinely seeks to understand the situation". This is what the Problem@Web project sought to do by focusing on the case of the SUB12 and SUB14 mathematics competitions. These online competitions have allowed youngsters in any suitable place, and at any suitable time, to engage themselves in tackling mathematical problems by utilising solving strategies with any digital tools that they have available.

In this book, our purpose is to provide a contribution to understanding the future of education through analysing the way that the digital generation tackles mathematical problems with the technologies of their choice at the time of their choice. This particular chapter provides an introduction to research relating to youngsters' mathematical problem-solving with technology, along with an overview of the Problem@Web project—a project that grew out of our interest in understanding how the youngsters of today tackle and solve moderately challenging mathematical problems using the digital tools of their choice.

1.2 Young People with Technology

In recent years there have been numerous discussions around the usage of digital technologies by today's young people. Parents and educators have sometimes expressed concern about the so-called digital natives' incessant use of technology (e.g. see Byron, 2008; Davies & Eynon, 2012). Worries about the amount of time that today's youngsters spend using digital technologies to play or communicate with friends are often voiced, and there is the concern that not only are young people dependent on technologies, but relentless use of technology means that they are socially inept and do not study, thereby causing conflict in families (Huisman, Edwards, & Catapano, 2012). Simultaneously, a range of authors and reports (such as Clark-Wilson, Oldknow, & Sutherland, 2011; Heiden, Fleischer, Richert, & Jeschke, 2011; OECD, 2012; Taborda, 2010) provide a different perspective, recognising that young people learn much through their everyday use of digital technologies.

Across the 21 countries that took part in the 2013 ICILS (Fraillon et al., 2014), the most extensive weekly use of software applications by Grade 8 students was "creating or editing documents" (28 % of students at least once a week) (p. 132). Use of most other applications was much less frequent, with "only 18 percent of the students… using education software designed to help with school study" (p. 251). Even so, Furlong and Davies (2012) found that many of the young people they studied had "access to a far wider range of digital resources to support their learning when working at home than in school or college" with these digital resources including "using their ICTs for organising, sharing and transporting their work (using personal or shared desktop folders, memory sticks); using electronic recommendation systems (e.g. Amazon, StumbleUpon); and publicly and commercially available learning-support materials (BBC Bitsize, Sparknotes)" (p. 55). What is more, Furlong and Davies found that if young people had "the necessary technical expertise to access a range of different technologies and platforms", and many of them did, along with "networking and collaborative skills", which again many of them had, then they could access "a far wider array of resources and ways of learning than is often available in school" (p. 60). As such, today's youngsters often acquire various competences through out-of-school experiences, including from extra school activities, free summer courses, when travelling with family or friends or via the Internet (Carreira, 2009).

These various considerations have prompted efforts by school systems to address the challenges raised by the increasing availability of various digital technologies. As a result, over recent years many countries have been making efforts to equip schools with new technologies (see Ranguelov, Horvath, Dalferth, & Noorani, 2011, for examples of this from across the European Union). Portugal, for instance, joined this movement through the Lisbon Strategy (March 2000) and the Lisbon Treaty (December 2007) which triggered a range of measures for technological development never seen before. The Educational Technological Plan enacted in Portugal enabled schools to be provided with cutting-edge equipment (desktop computers, laptops, wireless connection and interactive whiteboards). It enabled all schools in Portugal to have high-speed Internet connection, and it established a teacher training plan to certify teachers' technological competences.

This equipping of schools with computers and the Internet is paralleling the growth of digital technologies in the home. All this is enabling youngsters to access the world of online mathematics competitions such as SUB12 (for 10–12 years) and

SUB14 (13–14 years) described later in this chapter. These examples of mathematical problem-solving competitions take place at a distance through the Internet in a way that aims to attract the interest of today's youngsters. To find a mathematical problem on the Internet, to tackle it using the digital technologies at hand and to send a solution via the web (via online form or by e-mail) are some things that fit with the ways of being and communicating of today's young people. What is less clear is how the youngsters find effective and productive ways of thinking about mathematical problems and how they achieve a solution and communicate it using the digital resources that they have at hand. That is what the Problem@Web project sought to investigate.

1.3 Young People's Mathematical Problem-Solving with Technology

The work of Polya (1945, 1962) on mathematical problem-solving, taken to be the finding of a solution to a mathematical problem for which the solution strategy is not clear, inspired a very fruitful period of mathematics education research on the topic. Nowadays, mathematical problem-solving is featured as a central idea in defining the mathematics curriculum in many countries. Nevertheless, after this initial flourishing, the amount of research studies on mathematical problem-solving faded somewhat (English & Sriraman, 2010; Lester & Kehle, 2003; Lesh & Zawojewski, 2007). In this age of digital technology, there is a pressing need to renew the investigation of mathematical problem-solving. Being frequent users of new technologies and with likely extensive gaming and social media experiences, the characteristics of today's young people change the premises on which research investigations take place and require a new perspective.

At the same time, as is clear in recent initiatives promoted by the OECD (the Organisation for Economic Co-operation and Development), the demands of today's society are more and more in tune with problem-solving. This is borne out by problem-solving becoming an assessment domain in the OECD's Programme for International Student Assessment (PISA). For PISA 2012, for example, an assessment framework for problem-solving was devised and additional assessment methods implemented. What is more, the assessment of problem-solving for PISA 2012 was computer based allowing the real-time capture of students' capabilities, with student interactivity with the problem becoming "a central feature of the assessment" (OECD, 2013, p. 3).

The number of voices who are arguing for the use of computer technology in problem-solving, in any context, is increasing (Kim & Hannafin, 2011). These voices also turn to the distinct interaction that seems to exist between the problem-solver and the tools that the problem-solver may be using. This is especially in terms of the role of technologies as scaffolding resources, as expressed here:

1.3 Young People's Mathematical Problem-Solving with Technology

Problem solving involves situated, deliberate, learner-directed, and activity-oriented efforts to seek divergent solutions to authentic, personally meaningful problems through multiple interactions amongst problem solvers, tools, and related resources. (Kim & Hannafin, 2011, p. 404)

It is the nature of these interactions that are addressed by our Problem@Web research project as we study students solving mathematical problems whether in their home environment or in their classroom or other place where the computer is present, accessible and seen as useful by the solver.

The online nature of the qualifying rounds of the SUB12 and SUB14 competitions allows the possibility of reading each new proposed problem through the window of the computer (see Fig. 1.1 for a screenshot of the SUB12 and SUB14 homepage). This "window" is the immediate access to each new challenge but can also be the vehicle to start drawing a strategy of resolution, individually or with friends through social networks or e-mail. This variety of options seems to appeal to today's young people. The participants know the "rules of the game" and, therefore, know that to solve each problem correctly is not enough, but what is needed, above all, is to find ways of expressing their reasoning and make visible their strategies to the eyes of the organisers. In order to accomplish that, the participants know that they can use any of the digital tools that they have at their disposal and that they consider useful in their problem-solving. The requirement that the youngsters, in their digital submissions to the competition, provide a narrative of their problem-solving strategy, and of the path leading to their solution, reveals important forms of their exploratory and explanatory discourse closely aligned with mathematical thinking.

There is strong evidence that the use of digital technologies, mainly computers, changes the nature of the problem-solving activity (Lee & Hollebrands, 2006; Lopez-Real & Lee, 2006). One clear impact is that a mathematical question may no longer be so much a "problem" if the solver has a relevant computer tool available.

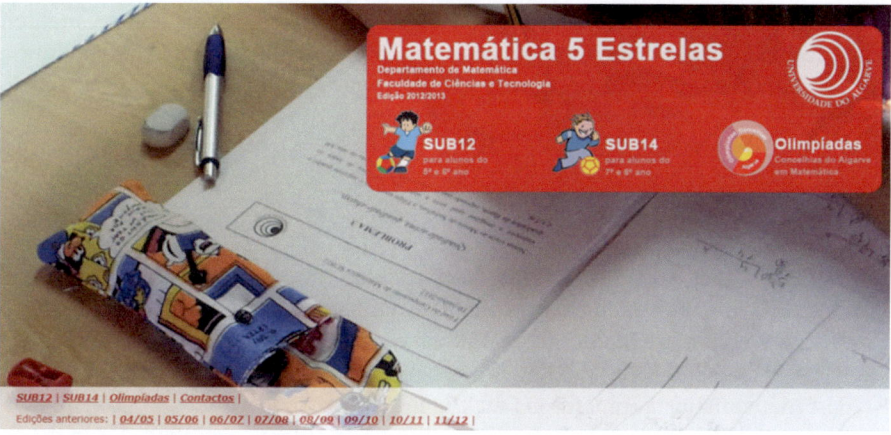

Fig. 1.1 The SUB12 and SUB14 homepage (edition 2012–2013)

This indicates that there is still much to learn about the connection between the use of technologies and the process of solving mathematical problems. Two research topics are the types of problem-solving approaches that might be possible through the use of digital technologies (such as visualisation, experimentation and simulation) and the different types of mathematical representations available and afforded by technologies that can be used to express and carry out mathematical thinking. The fact that young people familiar with digital technologies can learn and do mathematics in quite different ways from earlier generations reinforces the importance of studies into how today's youngsters tackle and solve mathematical problems. This is especially the case in out-of-school contexts in which the youngsters have the freedom to choose to use any digital technology that they have at hand.

Building on significant studies about mathematical problem-solving that were conducted during the twentieth century, research on mathematical problem-solving needs to find new ways to understand the nature of humans' approaches to mathematisable situations—especially as digital technologies are changing the approaches that youngsters are using to solve and communicate mathematically (Martinovic, Freiman, & Karadag, 2013). This entails acknowledging the existence of a new generation of learners—youngsters who are developing, mostly out of school, a set of competences that might help to underpin the skills and the sophistication required to learn beyond the school boundary. In the specific context of digital mathematical competitions (such as SUB12 and SUB14 described below), participants can communicate their reasoning about the mathematical problems in a creative way and can utilise any type of technological tool that they have at hand. This can add competences that, at times, school may neglect or overlook (Jacinto, Amado, & Carreira, 2009).

As Santos-Trigo (2004) argues, the use of technology can offer students an important window to observe and examine connections and relationships that become relevant during the process of solving a mathematical problem. The use of different tools, such as a spreadsheet or dynamic geometry software, offers youngsters the possibility of examining situations from perspectives that involve the use of various concepts and resources. Digital technologies also offer, as Jones, Geraniou and Tiropanis (2013) show, more possibilities for communication and collaboration during mathematical problem-solving.

1.4 The Research Focus

Informed by the overview of research relating to youngsters' mathematical problem-solving with technology, we now turn to providing an outline of the Problem@Web project including its rationale, aims and methods. The project grew out of our interest in understanding how youngsters participating in the SUB12 and SUB14 mathematics competitions tackle and solve what we term problems with "moderate" challenge (for two examples of such problems, see Figs. 1.2 and 1.3) and how they communicate their solutions—all mediated by the digital tools of their choice. An important aspect of these problems is that typically they are open

1.4 The Research Focus

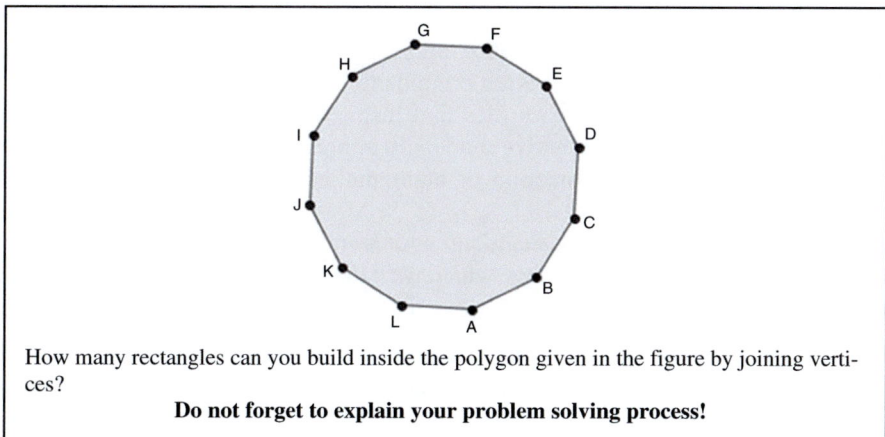

How many rectangles can you build inside the polygon given in the figure by joining vertices?

Do not forget to explain your problem solving process!

Fig. 1.2 Problem #3 of SUB12 (edition 2012–2013)

Four consecutive even numbers were distributed on the vertices of a tetrahedron. Each face of the tetrahedron was assigned the sum of the three numbers on its vertices. Then the numbers assigned to the faces of the tetrahedron were added and the total was 132. What were the numbers that were placed at the vertices of the tetrahedron?

Do not forget to explain your problem solving process!

Fig. 1.3 Problem #3 of SUB14 (edition 2010–2011)

to several alternative conceptual models serving as scaffolds for their resolution. At the same time, the multiple affordances of various digital tools are crucial to supporting thinking in terms of the different models that can lead to the solution, including models that young people would not likely produce without those tools. These affordances include imagery, pictorial and diagrammatic approaches supported by drawing and editing tools; tabular, numerical and relational algebraic thinking based on the spreadsheet tools; as well as measuring, constructing and manipulating geometrical figures with dynamic geometry software.

From the start of the competitions in 2005, we were surprised with solutions submitted by the young participants that showed a wealth of ways to represent and express mathematical reasoning. As educators we were motivated by the emergence of new ways of mathematical thinking and problem-solving by youngsters who engage in mathematical thinking via the computer. The solutions to the problems that we received over the years are a huge and varied database composed of images, textual written compositions, drawings, diagrams, tables, graphs, Excel, GSP or GeoGebra files that inspired us to try to understand and interpret the vastness of alternative models to approach a mathematical problem stemming from the interconnection of mathematical reasoning and digital fluency that was being revealed.

Our research project, *Mathematical Problem Solving: Perspectives on An Interactive Web-Based Competition*, which we call Problem@Web, had three main research foci:

(a) Strategies for solving mathematical problems, modes of representation and expression of mathematical thinking and the use of digital technologies in problem-solving
(b) Attitudes and emotions related to mathematics and mathematical problem-solving, both in school and beyond-school activities, considering students, parents and teachers
(c) Creativity expressed in mathematical problem-solving and its relation to the use of digital technologies

In this chapter, as with this whole book, we focus on the first of these; the thinking and strategies for tackling and solving mathematical problems, the modes of representation and expression of mathematical thinking of young participants and the uses they made of digital technologies in mathematical problem-solving.

1.5 The SUB12 and SUB14 Mathematics Competitions

The mathematical problem-solving competitions SUB12 and SUB14 are regional online competitions, covering the southern region of Portugal, that have been running annually since 2005. Organised by the Mathematics Department of the Faculty of Sciences and Technology of the University of Algarve, SUB12 aims at 5th and 6th graders (10–12-year-olds), while SUB14 addresses students in 7th and 8th grade (12–14-year-olds).

The two web-based competitions are presented on a single website (http://fctec.ualg.pt/matematica/5estrelas/), have similar rules and run in parallel on a yearly basis. Each competition entails two distinct phases: *the Qualifying* and *the Final*. The Qualifying phase progresses entirely online (using the competition website and through digital remote communication) by proposing a series of ten mathematical problems, one posted online each fortnight, while the Final is a half-day on-site contest held at the campus of the University of Algarve.

During the Qualifying phase, youngsters may choose to participate individually or in small teams of two or three. They send their answers to the problems by e-mail or through the electronic message editor available on the website; in either case, they can propose their solutions and attach any kind of files they wish. Our experience has been that many of the submitted answers are contained in files in Word,

1.5 The SUB12 and SUB14 Mathematics Competitions

Twelve colleagues agreed to have lunch together in a restaurant that was close to their office. Eight of the colleagues ordered the soup. Six of the colleagues ordered the mini beef. Four of the colleagues ordered the salad. All of them asked for some dish but none of them ordered exactly two dishes. How many colleagues asked for three dishes?

Do not forget to explain your problem solving process!

Fig. 1.4 Problem #7 of SUB14 (edition 2011–2012)

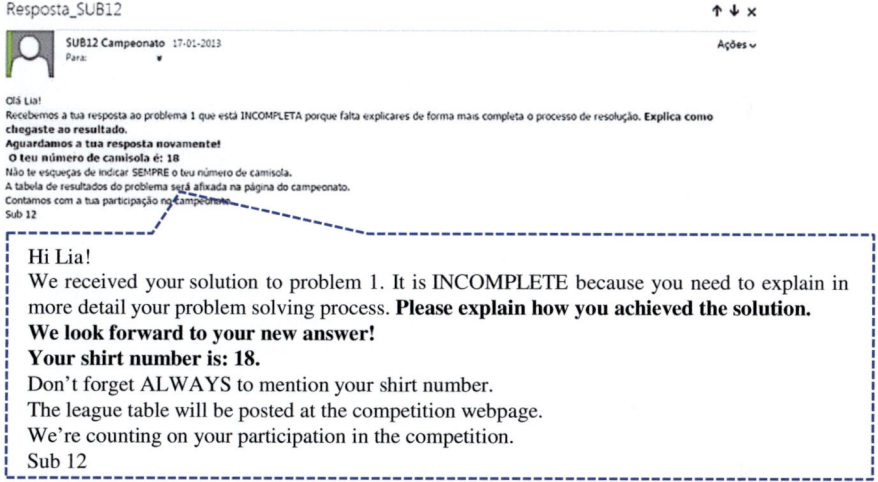

Fig. 1.5 Reply to a participant to their first submitted answer showing the assignment of a "shirt number"

Excel, Paint and GeoGebra or together with other means such as digital photos or scans of hand-written work.

Each problem proposed to participants in the Qualifying phase is displayed on the webpage and can be downloaded as a PDF file. As shown by the example problems in Figs 1.2, 1.3 and 1.4, every problem statement ends with the prompt "Do not forget to explain your problem-solving process!" as the request to include an explanation of each participant's thinking about the problem is a key rule of the competition. Students' answers that are submitted by e-mail or via the competition website are received in e-mail accounts specifically devoted to that purpose.

All messages sent by the participants receive a reply from the organisation of the competitions within 1 or 2 days. Figure 1.5 illustrates how the organisation acknowledges the submission of a solution to the first problem and welcomes the participant with an individual "shirt number" (like a player in a sports team, as per the competition logo in Fig. 1.1). The "shirt number" identifies the participant throughout the online Qualifying phase until the moment of the on-site Final.

The organising team, composed of senior mathematics teachers, continues to exchange e-mails with all the participants throughout the competition, offering formative and encouraging feedback, suggesting revisions when needed and offering hints to help in overcoming obstacles or just praising good answers and cheering the progress made. The feedback is matched to each case and to each solution, following the competition principles of being inclusive, constructive and motivating. Figures 1.5, 1.6 and 1.7 provide some examples of feedback sent to some participants in different situations and moments. The participants can also see the "league table" published on the website after each completed round. The table shows the records of correct, wrong, incomplete or absent answers to the problems posed since the beginning of the competition.

During each qualifying stage, students are allowed to submit revised solutions as many times as needed within the respective deadline, as the e-mail message in Fig. 1.8 reveals about a youngster who initially had trouble in solving one of the problems proposed by SUB14 involving relationships between sets. As students get regular feedback, they are prompted to present their ways of thinking about the problem in their own words and by whatever means they choose, or else to elaborate

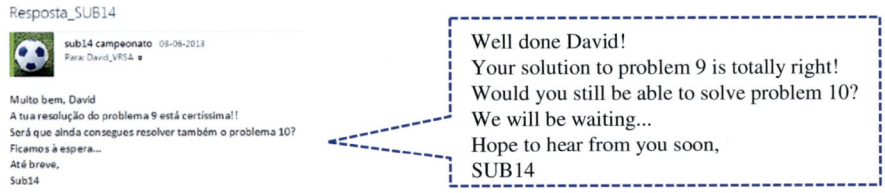

Fig. 1.6 Feedback to a participant with one well-explained solution and one still to do

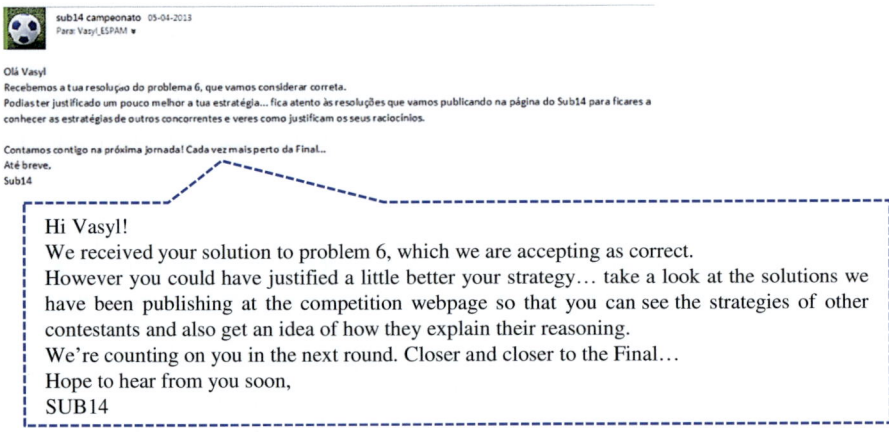

Fig. 1.7 Feedback to a participant with a satisfactory solution and a recommendation to improve the explanation given

1.5 The SUB12 and SUB14 Mathematics Competitions

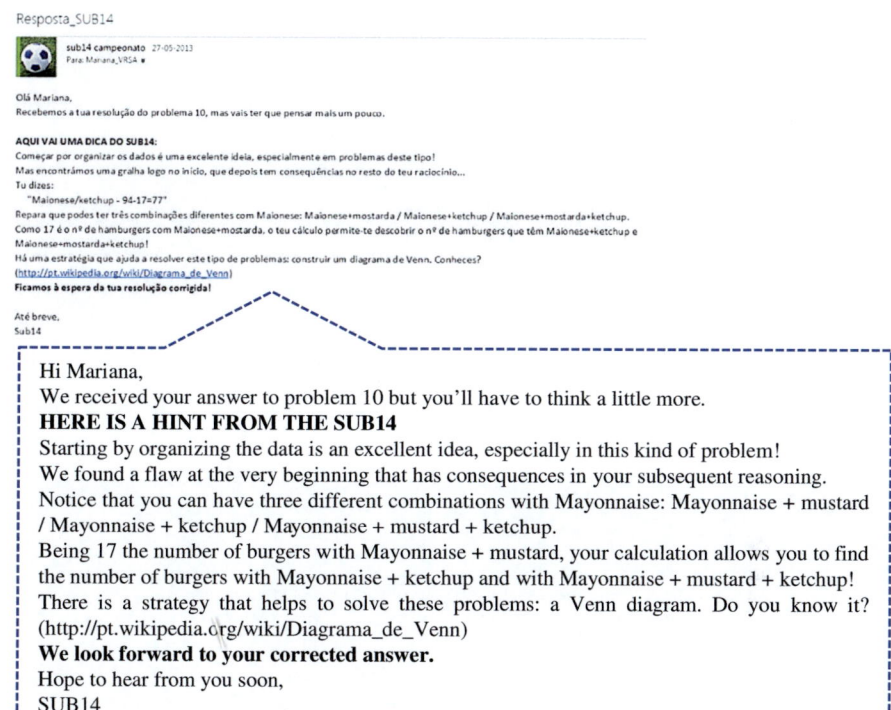

Fig. 1.8 Feedback to a participant with an unsatisfactory solution and some clues to help in solving the problem

more on their solutions. The absence of a convincing and clear explanation of the solution achieved means an incomplete or faulty answer to the problem.

The number of students participating in the two competitions has grown over the years, and recently, it has reached almost 2000 children in SUB12 and 800 in SUB14. As the Qualifying phase progresses, it is usual to notice a fall in the number of persisting competitors, mostly caused by attrition. This is more pronounced around the middle of the competition. Usually, about 10–15 % of the initial number of participants reaches the Final phase.

The Final is a half-day on-site contest held at the campus of the University of Algarve where the finalists, their families and teachers are welcomed. At the Final, the young competitors are each given a set of five problems to be solved in 2 h (Fig. 1.9a, b). In the meantime, parents, teachers and other accompanying guests attend a workshop, a seminar or other forms of interactive activities related to mathematics and mathematical problem-solving. The Final culminates with the award ceremonies for the three winners of each of the two competitions (Fig. 1.10a, b).

Throughout the history of this competition, a number of distinctive characteristics have stood out: (1) it proposes non-routine mathematical problems, usually allowing several ways to be solved (see two more examples in Figs. 1.11 and 1.12);

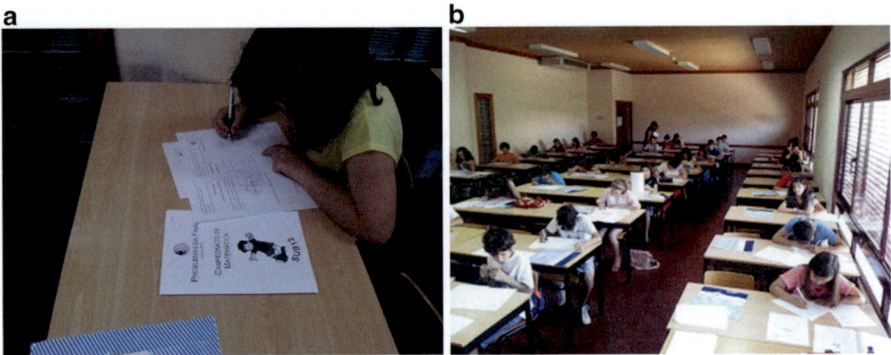

Fig. 1.9 Photos of the final ((**a**) student solving the problems; (**b**) one of the rooms at the final)

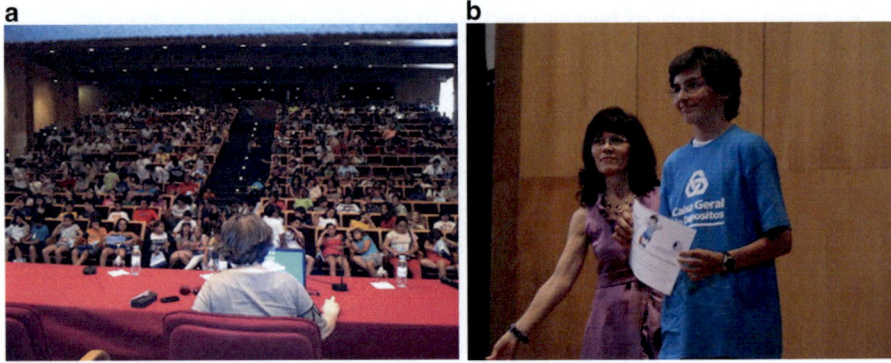

Fig. 1.10 Photos of the final ((**a**) the awarding ceremony; (**b**) the first prize awarding)

(2) problems are not intended to fit any particular school curricular topic; (3) the main trend is on moderate mathematical challenges; (4) the competition explicitly requires participants to disclose the process of finding the solution; (5) it is close to teachers and families in the sense that it encourages their support of the young participants (see Fig. 1.13); (6) opportunities for reformulating and resubmitting answers are offered to all participants; (7) all types of media used to find and develop solutions are welcome (see Fig. 1.14); (8) communication and interaction are carried out through digital web-based and e-mail infrastructures; (9) interesting and diverse proposed solutions are published on the competition website; and (10) the competitive component is concentrated within the Final phase of the competition rather than throughout.

Up to this point, we have displayed a few of the many problems posed in the two competitions, always different from edition to edition. Throughout the book, we return to some of them and bring in several more, but what interests us, above all, is to discuss and dissect a variety of solutions produced by young participants that reveal their ingenious ways of solving the problems and conveying their mathematical reasoning through the digital tools they choose to use. The example in Fig. 1.14

1.5 The SUB12 and SUB14 Mathematics Competitions

Two identical coffee pots are full of *latte* (coffee with milk). In the blue coffee pot, 3/5 of the *latte* is milk and the rest is coffee. In the brown coffee pot, 3/4 of the *latte* is milk and the rest is coffee. From the blue coffee pot, half of the *latte* was consumed. Then this coffee pot was refilled using the *latte* that was in the brown coffee pot. After that, what percentage of coffee was in the *latte* in the blue coffee pot?

Do not forget to explain your problem solving process!

Fig. 1.11 Problem #6 of SUB12 (edition 2012–2013)

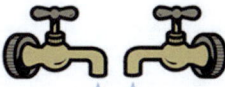

Mr. Boniface has a tank in his garden that needs to be filled regularly and he may use two taps with different flow rates. One of the taps fills the tank in 6 hours and the other tap fills the same tank in 3 hours. Early in the morning, Mr. Boniface saw that the tank was empty and opened the first tap (which pours less). When the tank was half of its capacity he also decided to open the second tap (which pours more). How long did it take to fill the tank, since he opened the first tap?

Do not forget to explain your problem solving process!

Fig. 1.12 Problem #3 of SUB14 (edition 2011–2012)

Fig. 1.13 A snapshot of the competition website, showing an informative post about help seeking

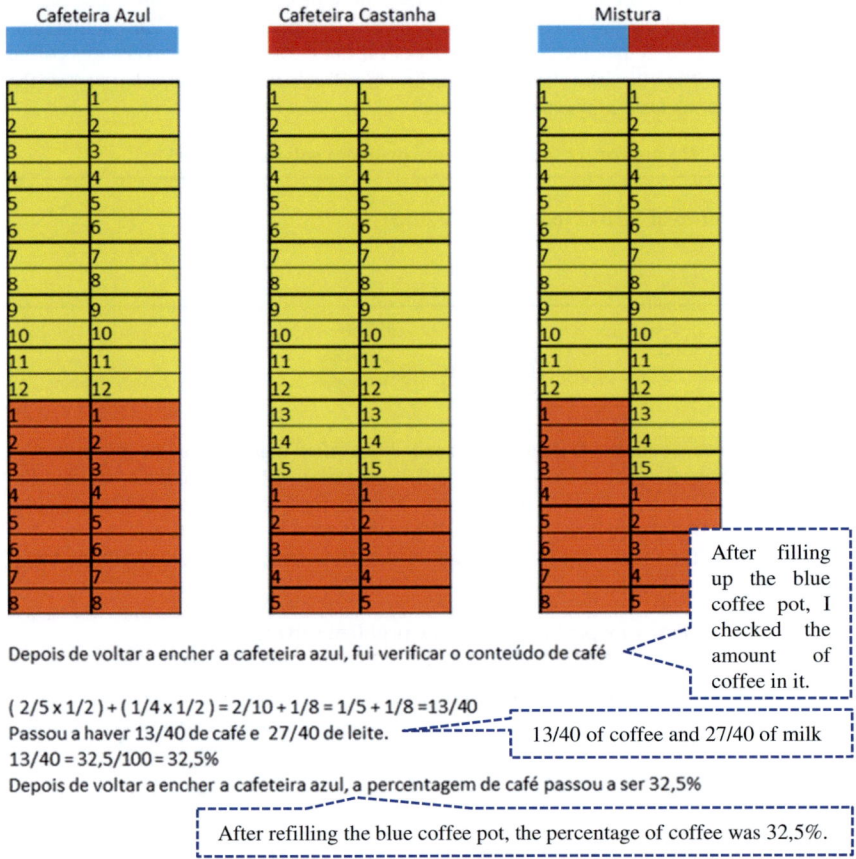

Fig. 1.14 A creative solution in a Word file to the problem of the coffee pots (Problem #6 of SUB12, edition 2012–2013)

illustrates that an interesting solution can be produced with common digital tools (see the "coffee pots" problem in Fig. 1.11). The solution consisted of a Word file used to create a schematic representation of the content of each pot, based on two tables and on colours that represent the coffee and the milk and on a third table that draws on the previous ones to illustrate the ratio of coffee and milk after the mixing.

1.6 Methodological Issues

One of the primary goals of the Problem@Web project was to understand how today's young people solve moderately challenging mathematical problems using the technology of their choice. Given the context of the SUB12 and SUB14 competitions, and our overall research objectives, we chose to conduct the major part of our research using a qualitative methodology. In terms of our qualitative data sources, we interviewed many young participants, as well as some former participants, several mathematics teachers who followed the competitions and also youngsters' parents and relatives; we collected all the e-mails and answers submitted throughout three editions of the competitions, and other informative data was obtained through observation, photos and videos, namely, during the Finals.

To complement the qualitative methodology, we found it important to accompany our qualitative data with some quantitative and statistical data aimed at providing a multi-dimensional characterisation of the people involved in our study: the youngsters who are participating in the competitions and their preferences, experiences and views on technologies and problem-solving. We therefore carried out a survey, after the 2011–2012 edition of the competitions, by supplying an online questionnaire at the end of the school year to all students of 5th to 8th grade from the Algarve region of southern Portugal, for which the project was assisted by the Algarve Regional Department of Education. The response rate of the respondents who participated in the competitions on the 2011–2012 edition was close to 20 %, corresponding to a total of $n=350$ individuals.

The questionnaire was composed of 25 closed questions, divided into four sections: the youngsters' relationship with technologies, their relationship with mathematics, their relationship with problem-solving and a final section only for those involved in the competitions on their views on participating in SUB12 or SUB14. These quantitative data were processed through various statistical methods from descriptive statistics to multivariate statistical analysis. The analysis of these data allowed the generation of interesting results when combined and coordinated with the qualitative data, in particular from the interviews and observations we conducted. We also expected that these data would provide useful information to help the characterisation of the youngsters involved in the competitions, in terms of the experience they had with a number of home digital tools that are today widely available and disseminated.

As qualitative research is necessarily concerned with understanding symbolic material, it requires some degree of interpretation (Delamont, 2012). Qualitative content analysis (Mayring, 2004; Schreier, 2012) is particularly appropriate when dealing with rich and authentic empirical data that can be addressed by selecting theoretical concepts and theory-driven analytical tools against which the reading and interpretation of the material is produced. In adopting this method, we do not change the materials selected but we understand them in the context where they originated and search for the meaning they bring to the questions and problems under study. So it is important to note that the context in which this research takes place cannot be ignored; on the contrary, in qualitative research, the data collected cannot be independent of the context of research.

It is precisely the context in which these competitions develop, with their own and distinctive features, that makes them meaningful as a research context and the empirical data rich as information sources. Thus, we consider that this research takes on a naturalistic design in that it analyses a phenomenon in a particular context that is significant from the point of view of purpose, function, institutional framing, social and educational setting, etc., in which, as researchers, we choose not to interfere regardless of our proximity to the competition environment (Schreier, 2012).

Even so, we know from accounts from the teachers of youngsters participating in the SUB12 and SUB14 competitions that the mathematical problems published on the website were widely used in schools, both in mathematics classes and in other school contexts such as libraries, mathematics after-school clubs or supervised study classes. The recognition by teachers of the importance of the resource provided by the competition website was particularly important for the research team as this provided the opportunity to extend our research to the school context.

Thus, it was possible to complement the information gathered in the online and beyond-school competitions, with information from classrooms, thereby allowing us to develop a more rounded understanding of the phenomenon of youngsters solving non-routine mathematical problems with technology in the school context. This meant, for example, that we could obtain data from an 8th grade class (of 13–14-year-olds) where students spent a number of lessons tackling various problems given in SUB14. Here the research took more of the form of a teaching experiment (Steffe & Thompson, 2000). The knowledge gained through the data collected in the classroom was particularly important for our project because we were not able to observe youngsters within the online phase of the competition. As such, data collection in the school setting provided important information to help us understand how youngsters think and act on these problems in school where the use of digital tools was available to them.

As each edition of the qualifying phase involved ten problems on various mathematical topics (such as geometry, algebra and number, combinatorial and logical reasoning), the number of digital documents collected over the years, including e-mails and attached files, became increasingly large. To select the data to be analysed, it was necessary to establish some criteria. One of these was the possibility of comparing students' productions in the competition with their efforts in the school context.

To develop our data analysis, we chose problems from different mathematical topics. These can be considered, in a sense, as a trilogy, as they address the major concepts of invariance, quantity variation and co-variation. This choice allowed us to give a very broad idea of the wealth of possibilities that technology use can provide in problem-solving and in problem-driven conceptual development.

The process of analysing these data was a process of constructing meaning for the youngsters' productions gathered in various forms such as images produced in PowerPoint or Word files, or constructions in GeoGebra, or tables and formulas in Excel. This was a complex process that usually involved a sequence of stages.

The encoding of the data is extremely important in qualitative content analysis, where coding is usually considered as a conceptual device. Some encoding techniques involve integrating data-driven codes with theory-driven ones (Fereday & Muir-Cochrane, 2006). Conducting a data-driven encoding entails a form of pattern recognition within the data, where emerging themes become the categories for analysis. In contrast, theory-driven coding is based on a preliminary set of categories developed a priori, based on the research question and the theoretical framework. In our research we followed this combined approach at various times. For example, looking for patterns and themes from a set of solutions to a problem involving quantity variation led to ways of ranking different spreadsheet-based solutions revealed in those solutions; in addition, we used theoretical concepts, like the idea of co-action between the solver and the tool, to complement the first encoding and generate evidence of the relationship between the resolution of the problem and the use of Excel, thus reaching a second level of interpretive understanding (Fereday & Muir-Cochrane, 2006). Schreier (2012) considers that encoding is a device that helps us think through our concepts by reference to our data, and as such, the process creates a trail of evidence that legitimates and corroborates the codes. In this way, conceptual coding involves creating links between data and concepts, between concepts and between data. This is how, by looking at our data from new and different angles, we discovered new aspects and new ways to question and discuss the data. It is through this process that we have sought to generate analytical theory and extend pre-existing theory from our data.

1.7 Concluding Comments

In this chapter we have provided an introduction to research relating to youngsters' mathematical problem-solving with technology, together with an overview of the Problem@Web project including its rationale, aims and methods. The research project has its origin in the steady-growing evidence from earlier editions of the competitions that the SUB12 and SUB14 problem-solving competitions are unique and highly valuable contexts in which to conduct studies on the ways students use, as they decide, the digital resources they have at their disposal to solve moderately challenging problems with their own strategies and mathematical representations.

So far, research has not been able to say in detail exactly what difference the use of technology can make in effective learning. There are results suggesting that it is *how* the technology is used that is more important than *what* technology is used. In taking forwards the research, Higgins, Xiao and Katsipataki (2012) consider that it is vital to understand better what makes a particular use of technology really decisive for the success of learning, arguing that it is important to look closely at promising uses of technology by youngsters:

> We need to know more about where and how it [the technology] is used to greatest effect, then investigate to see if this information can be used to help improve learning in other contexts. We do not know if it is the use of technology that is making the difference. (Higgins et al., 2012, p. 3)

Importantly, we must not forget that the use of technology in learning, particularly mathematics, is no longer confined to school, and this means that we also need to learn more about how today's youth develops technological skills and problem-solving skills in unexpected and often unknown ways outside school. In our project we embraced this goal: to scrutinise and analyse numerous solutions to mathematical problems solved with the mediation of some digital tool, in looking at the strategies, forms of expression and capability to take advantage of such tools that many youngsters are demonstrating.

It is not about terms such as digital natives or equivalents; such labels correspond to a generic, imperfect and possibly sketchy idea of the new generation. What really matters is to see how the experience of today's youngsters might illuminate our knowledge of the productive relationship they can establish between digital technologies (that are certainly appealing to many of them) and their successful mathematical problem-solving. In agreeing that "just because young people have grown up with technology it does not mean they are experts in its use for their own learning" (Higgins et al., 2012, p. 20), we think it is important to find ways to unveil youngsters' powerful conceptual models and effective ways of expressing mathematical thinking on non-routine problems through the use of commonly available digital tools.

The mathematical competitions SUB12 and SUB14 are not only an online problem-solving environment and are not just about digital technologies. We have described the competitions by a set of particular characteristics that are in line with supportive and inclusive learning environments. Communication with the competitors is conducted remotely yet aims at being supportive, helpful and encouraging. The mathematical problems posed during the competitions are, in any case, a challenge to young people who are willing to experiment and make use of commonly used digital tools, such as Excel, or GeoGebra, or Word, or Paint.

The next chapter gives a first idea of who these young people involved in the competitions are (whether fans or not of digital technologies), what they claim to know about the use of digital tools and what they say about their use in mathematical problem-solving in this beyond-school context. This is one component of what we offer in this book of a "considered and rigorous investigation that includes the perspectives of young people and their teachers, and genuinely seeks to understand the situation" (Bennett et al., 2008, p. 784).

References

Bennett, S., Maton, K., & Kervin, L. (2008). The "digital natives" debate: A critical review of the evidence. *British Journal of Educational Technology, 39*(5), 775–786. http://dx.doi.org/10.1111/j.1467-8535.2007.00793.x.

Byron, T. (2008). *Safer children in a digital world: The report of the Byron Review*. London: Department for Children, Schools and Families, and the Department for Culture, Media and Sport. http://www.education.gov.uk/publications/eOrderingDownload/DCSF-00334-2008.pdf.

Carreira, S. (2009). Matemática e tecnologias: ao encontro dos "nativos digitais" com os "manipulativos virtuais". *Quadrante, XVIII*(1-2), 53–85.

References

Clark-Wilson, A., Oldknow, A., & Sutherland, R. (2011). *Digital technologies and mathematics education*. London, UK: Joint Mathematical Council of the United Kingdom. http://cme.open.ac.uk/cme/JMC/Digital%20Technologies%20files/JMC_Digital_Technologies_Report_2011.pdf.

Davies, C., & Eynon, R. (2012). *Teenagers and technology*. London: Routledge.

Delamont, S. (Ed.). (2012). *Handbook of qualitative research in education*. Cheltenham: Edward Elgar Ltd.

English, L., & Sriraman, B. (2010). Problem solving for the 21st century. In B. Sriraman & L. English (Eds.), *Theories of mathematics education: Seeking new frontiers* (pp. 263–295). New York, NY: Springer. http://dx.doi.org/10.1007/978-3-642-00742-2_27.

Fereday, J., & Muir-Cochrane, E. (2006). Demonstrating rigor using thematic analysis: A hybrid approach of inductive and deductive coding and theme development. *International Journal of Qualitative Methods, 5*(1), 80–92. http://ejournals.library.ualberta.ca/index.php/IJQM/article/view/4411/3530.

Fraillon, J., Ainley, J., Schulz, W., Friedman, T., & Gebhardt, E. (2014). *Preparing for life in a digital age: The IEA International Computer and Information Literacy Study International Report*. Heidelberg: Springer. doi:10.1007/978-3-319-14222-7.

Furlong, J., & Davies, C. (2012). Young people, new technologies and learning at home: Taking context seriously. *Oxford Review of Education, 38*(1), 45–62. http://dx.doi.org/10.1080/03054985.2011.577944.

Heiden, B., Fleischer, S., Richert, A., & Jeschke, S. (2011). Theory of digital natives in the light of current and future e-learning concepts. *International Journal of Emerging Technologies in Learning (iJET), 6*(2), 37–41. http://dx.doi.org/10.1007/978-3-642-33389-7_38.

Helsper, E., & Eynon, R. (2010). Digital natives: Where is the evidence? *British Educational Research Journal, 36*(3), 503–520. http://dx.doi.org/0.1080/01411920902989227.

Higgins, S., Xiao, Z., & Katsipataki, M. (2012). *The impact of digital technology on learning: A Summary for the education endowment foundation*. Durham, UK: Durham University. http://educationendowmentfoundation.org.uk/uploads/pdf/The_Impact_of_Digital_Technologies_on_Learning_FULL_REPORT_(2012).pdf.

Huisman, S., Edwards, A., & Catapano, S. (2012). The impact of technology on families. *International Journal of Education and Psychology in the Community, 2*(1), 44–62.

Jacinto, H., Amado, N., & Carreira, S. (2009). Internet and mathematical activity within the frame of "Sub14". In V. Durand-Guerrier, S. Soury-Lavergne & F. Arzarello (Eds.). *Proceedings of the Sixth Congress of the European Society for Research in Mathematics Education* (pp. 1221–1230). Lyon, France: Institut National de Recherche Pédagogique. http://ife.ens-lyon.fr/editions/editions-electroniques/cerme6/.

Jones, K., Geraniou, E., & Tiropanis, T. (2013). Patterns of collaboration: Towards learning mathematics in the era of the semantic web. In D. Martinovic, V. Freiman, & Z. Karadag (Eds.), *Visual mathematics and cyberlearning* (pp. 1–21). Dordrecht: Springer. http://dx.doi.org/10.1007/978-94-007-2321-4_1.

Kim, M. C., & Hannafin, M. J. (2011). Scaffolding problem solving in technology-enhanced learning environments (TELEs): Bridging research and theory with practice. *Computers and Education, 56*(2), 403–417. http://dx.doi.org/10.1016/j.compedu.2010.08.024.

Lee, H. S., & Hollebrands, K. F. (2006). Students' use of technological features while solving a mathematics problem. *Journal of Mathematical Behavior, 25*(3), 252–266. http://dx.doi.org/10.1016/j.jmathb.2006.09.005.

Lesh, R., & Zawojewski, J. (2007). Problem solving and modeling. In F. K. Lester (Ed.), *Second handbook of research on mathematics teaching and learning* (pp. 763–804). Charlotte, NC: Information Age Publishing.

Lester, F. K., & Kehle, P. E. (2003). From problem solving to modeling: the evolution of thinking about research on complex mathematical activity. In R. Lesh & H. M. Doerr (Eds.), *Beyond constructivism: Models and modeling perspectives on mathematical problem solving, learning, and teaching* (pp. 501–517). Mahwah, NJ: Lawrence Erlbaum Associates.

Lopez-Real, F., & Lee, A. M. S. (2006). Encouraging the use of technology in problem solving: Some examples from an initial teacher education programme. *International Journal for Technology in Mathematics Education, 13*(1), 23–29.

Martinovic, D., Freiman, V., & Karadag, Z. (2013). *Visual mathematics and cyberlearning*. Dordrecht, The Netherlands: Springer. http://dx.doi.org/10.1007/978-94-007-2321-4.

Mayring, P. (2004). Qualitative content analysis. In U. Flick, E. Von Kardoff, & I. Steinke (Eds.), *A companion to qualitative research* (pp. 266–269). London: Sage.

OECD. (2012). *Connected minds: Technology and today's learners*. Paris: OECD Publishing. http://dx.doi.org/10.1787/9789264111011-en.

OECD. (2013). *PISA 2012 assessment and analytical framework: Mathematics, reading, science, problem solving and financial literacy*. Paris: OECD Publishing. http://dx.doi.org/10.1787/9789264190511-en.

Polya, G. (1945). *How to solve it: A new aspect of mathematical method*. Princeton, NJ: Princeton University Press. Reprinted 1957, Doubleday, Garden City, NY.

Polya, G. (1962). *Mathematical discovery: On understanding, learning, and teaching problem solving* (Vol. I, II). New York, NY: Wiley.

Prensky, M. (2001). Digital natives, digital immigrants. *On the Horizon, 9*(5), October, (n/p.). NCB University Press.

Prensky, M. (2006). *Don't bother me, Mom, I'm learning! How computer and video games are preparing your kids for 21st century success and how you can help!* St. Paul, MN: Paragon House.

Ranguelov, S., Horvath, A., Dalferth, S., & Noorani, S. (2011). *Key data on learning and innovation through ICT at school in Europe 2011*. Brussels: Education, Audiovisual and Culture Executive Agency, European Commission. Retrieved November 26, 2014, from http://eacea.ec.europa.eu/education/eurydice/documents/key_data_series/129EN.pdf.

Roberts, D. F., & Foehr, U. J. (2005). *Generation M: Media in the lives of 8–18 year-olds*. Menlo Park, CA: The Henry J. Kaiser Family Foundation. Retrieved November 26, 2014, from http://kff.org/other/generation-m-media-in-the-lives-of/.

Santos-Trigo, M. (2004). The role of technology in students' conceptual constructions in a sample case of problem solving. *Focus on Learning Problems in Mathematics, 26*(2), 1–17.

Schreier, M. (2012). *Qualitative content analysis in practice*. London: Sage.

Steffe, L. P., & Thompson, P. W. (2000). Teaching experiment methodology: Underlying principles and essential elements. In R. Lesh & A. E. Kelly (Eds.), *Research on design in mathematics and science education* (pp. 267–307). Hillsdale, NJ: Lawrence Erlbaum Associates.

Taborda, M. J. (Ed.). (2010). *Nativos Digitais portugueses: Idade, experiência e esferas de utilização das TIC*. Lisbon, Portugal: OberCom (Observatório da Comunicação).

Tapscott, D. (1998). *Growing up digital: The rise of the Net Generation*. New York: McGraw-Hill.

Tapscott, D. (2009). *Grown up digital: How the Net Generation is changing your world*. New York: McGraw-Hill.

Chapter 2
Youngsters Solving Mathematical Problems with Technology: Their Experiences and Productions

Abstract Over several years of the SUB12 and SUB14 online mathematical competitions, we became aware of the technological fluency of many of the young participants. We draw on quantitative data from a survey that was administered online by inviting all participants to respond. The data show how the participants describe themselves in terms of their experience with several digital tools. We have found that they feel comfortable with the use of text and presentation editors and know how to use several tools for writing, creating tables and constructing diagrams and visual representations. In contrast, they seem to be less capable with spreadsheets (especially as a mathematical tool) and dynamic geometry software. Some participants preferred to submit copies of their hand-written answers to the problems as scanned images or digital photos. In reporting the results of our survey, we present a selection of solutions covering a palette of examples that help to exemplify the skills and fluency of the competition participants. They unveil a particular trait of this mathematical problem-solving activity since these digital solutions bring together problem-solving and the expressing of mathematical thinking.

Keywords Digital technologies • Mathematical texts • Mathematical tables • Mathematical images and diagrams • Spreadsheets • Dynamic geometry software • Technological fluency

2.1 Introduction

Our experience in the Problem@Web project has been that throughout the various rounds of the web-based mathematical competitions SUB12 and SUB14, the young participants revealed their digital fluency in the use of home digital technologies to solve the problems that were posed during the online phase. Many of the solutions submitted by the participants showed use not only of general tools such as text editors, spreadsheets, presentation editors, image processing tools, file converters, scanning hardware, digital cameras, smartphones and online repositories, but also specialised tools such as dynamic geometry software and programming languages.

We became aware that while this technological fluency was clearly applicable across various uses, it had some important differences within the universe of the young participants. This led us to investigate the following question: who are these youngsters that are engaging in these online mathematical competitions?

In this chapter, we provide a picture of those youngsters who participated in the SUB12 and SUB14 mathematical competitions during the lifetime of the Problem@Web project. This picture is constructed from an analysis of data collected through an online survey of the participants, interviews with some of the youngsters, solutions submitted during the competitions and e-mail messages exchanged between the youngsters and the organisation. The questionnaire consisted of two parts, the first of which was given to all school children in grades 5–8 in the Algarve, whether or not they were participating in the competitions. In this first part, general descriptive data were gathered: gender, age, school and grade and attainment level in mathematics. Following this there were questions about the youngsters' relationships with technology, both at home and at school, and with mathematics and mathematical problem-solving. The second part of the questionnaire was completed only by youngsters who had participated in at least one of the competitions; it focused on their participation. Most questions in this second part were Likert-type questions answered on a scale with four or five levels. The number of respondents was $n = 350$, corresponding to a return rate of around 20 % (see Chap. 1, Sect. 1.6 for more details of the questionnaire methodology). We performed a descriptive statistical analysis in order to obtain a general characterisation of the sample in terms of their familiarity with technology use. We supplement this statistical analysis with an analysis of a sample of submitted solutions to problems in order to illustrate the youngsters' ways of using various digital tools.

To begin their participation in the competitions, youngsters only have to send their solution to Problem #1, indicating their full name, grade and class and the name of the school and its location. No subscription is necessary to become enrolled in the competition and continue to send the solutions to the problems. To submit an answer to a problem, the participants can choose to use an online form provided at the competition webpage or they can use their personal e-mail as long as they include all the requested identification data. The online form available at the competition webpage (Fig. 2.1) provides a simple text editor for the inclusion of the solution to the problem and allows the attaching of files. Participation may be individual or in groups of two or three youngsters, preferably of the same grade and class.

2.2 The Participants in the Mathematical Competitions SUB12 and SUB14

The web-based mathematical competitions SUB12 and SUB14 are primarily aimed at youngsters in Portugal and specifically from the south region of the country, comprising the Algarve and the Alentejo. Having been run annually since 2005, the number of youngsters participating in the two competitions grew over the years.

2.2 The Participants in the Mathematical Competitions SUB12 and SUB14

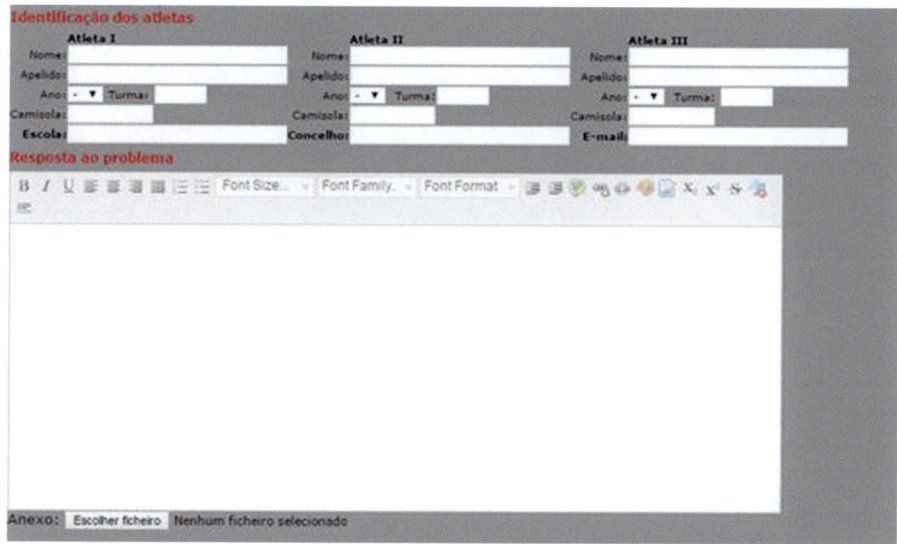

Fig. 2.1 Image of the online form available at the competition webpage for submitting a solution

By 2013 it reached about 2000 participants per year in SUB12 and 800 per year in SUB14, roughly 60 % from the Algarve and 40 % from the Alentejo. The contrast between SUB12 and SUB14 could be due to several factors, possibly the result of an increase in the number of school subjects in the transition from Grades 5–6 (the 2nd cycle of basic education in Portugal) to Grades 7–9 (the 3rd cycle). More school subjects involve more study and naturally less time to participate in extracurricular activities. The e-mail database from the competitions shows that the participants sometimes themselves justified a delay in sending their answers to the competitions due to school assessment tests. Another reason for the differing participation could be that youngsters are finding new interests and new activities as they grow older.

Relevant statistics for Portugal (PORDATA, October 2013) show that over the past decade, the number of Portuguese male youngsters aged 10–14 was slightly higher than the number of Portuguese female youngsters of the same age, with the respective percentages in 2012 being 52 % boys and 48 % girls. Our data about the participants in SUB12 and SUB14 show a slightly different demography: nearly 55 % of the respondents were girls and about 45 % were boys (Fig. 2.2). Throughout the various editions of both competitions, the number of girls was always higher than the number of boys.

The ages of our survey respondents ranged between 10 and 15 years old (Fig. 2.3). This might be anticipated given that SUB12 aims at 5th and 6th graders (10–12-year-olds), while SUB14 addresses students in 7th and 8th grade (12–14-year-olds). This data is consistent with the fact that the youngsters begin participating in Grade 5 at about 10 years of age.

The 11-year-olds seemed to be the most enthusiastic for these competitions. As Fig. 2.3 shows, about 33 % of respondents were aged 11 and were attending 5th or

Fig. 2.2 Gender distribution of the respondents to the SUB12 and SUB14 survey

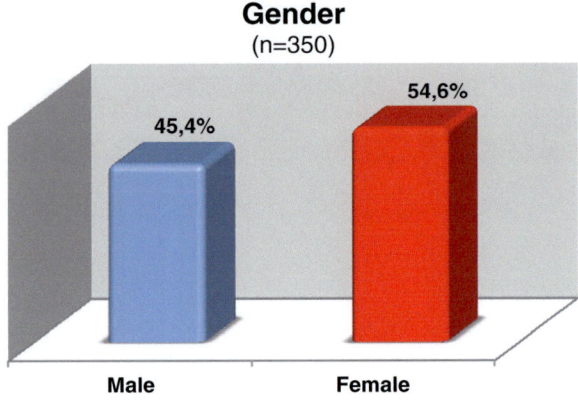

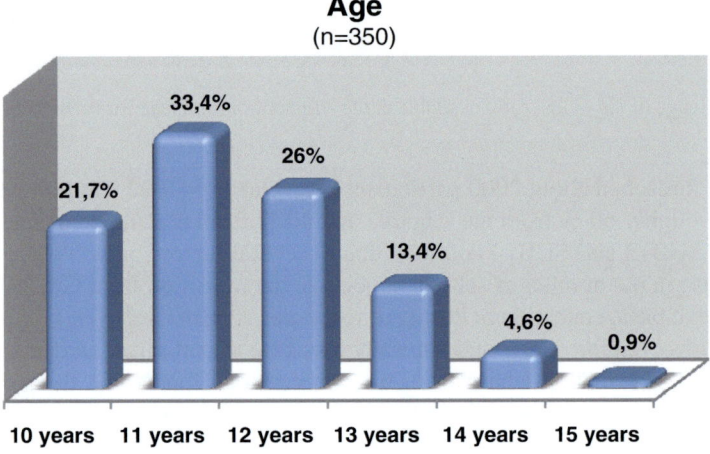

Fig. 2.3 Age distribution of the respondents to the SUB12 and SUB14 survey

6th grade. They may be participating for the first or second time. Beyond the age of 12 years, there is a decrease in the number of respondents. The percentage of respondents decreases as the school grade increases (Fig. 2.4). Youngsters attending the 5th grade were the most represented. The reasons for this enthusiasm may be varied; using the computer and the Internet is likely to be one of the aspects that motivated the younger children, as some participants mentioned when interviewed. For example, Rui, a young boy who participated in the competitions over a 4-year period, two in SUB12 and two in SUB14, stated in an interview:

> Rui: It was something different, that of being over the Internet. I enjoyed using the computer.

It is also interesting to consider the school mathematics attainment of youngsters who participate in these competitions. The inclusive character of the SUB12 and SUB14 competitions, with freedom on how to obtain solutions and express them,

2.2 The Participants in the Mathematical Competitions SUB12 and SUB14

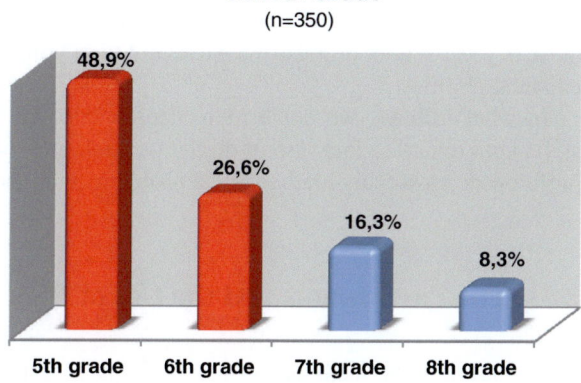

Fig. 2.4 School grade distribution of the respondents to the survey

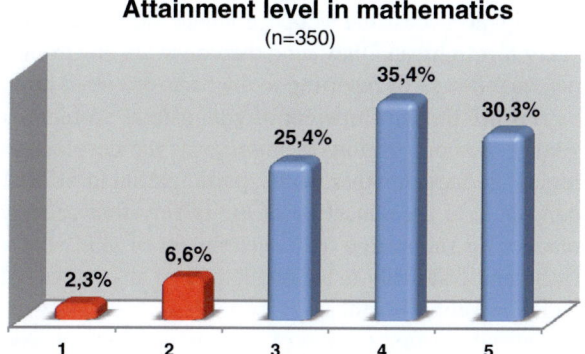

Fig. 2.5 Mathematics attainment of the respondents to the survey

the constructive feedback and the extended time that each participant has to solve the problems including correcting any faults, may be some of the aspects that contribute to their wide scope of interest. As is evident in Fig. 2.5, the competitions do not just attract the most talented and the most successful youngsters in school mathematics. On a scale of attainment from 1 to 5 (with 5 being the highest), there was a considerable proportion of youngsters who may be considered as average or low attaining in mathematics (around 34 % were at level 3 or lower).

Thus, we have a first overall sense of who these youngsters, interested in participating in moderately challenging mathematical problem-solving competitions taking place in an online environment, appear to be. They are just about evenly distributed by gender, although with a small prevalence of female participants. The contestants who seem to more enthusiastically adhere to the competitions are attending 5th and 6th grades, therefore being engaged in SUB12 and being also the youngest ones. In addition, it seems plausible that this kind of environment which encourages the use of the computer and the Internet is motivating for the youngest ones, who may see in it an opportunity to get in touch with digital technologies. Finally, we may say that the youngsters involved in these competitions, whether or not they complete the entire Qualifying phase and get selected for the Final, are heterogeneous in terms of their school mathematics attainment, which tells us that these beyond-school projects meet a diversity of youngsters in regard to their mathematical abilities.

In what follows, we delve more deeply into the youngsters' capabilities and skills with regard to their use of digital technologies, acknowledging their different preferences on solving mathematical problems with technologies.

2.3 The Participants and the Use of Digital Technologies

Participation in the mathematics competitions SUB12 and SUB14 requires the use of a computer with Internet connection to access the website. The existence of these technological resources is indispensable to get engaged in the competitions; the participants access the webpage where they find each new problem that is launched every fortnight, and they send their answers electronically as well, either using their personal e-mail or resorting to the tools provided by the website.

Besides the development of youngsters' problem-solving competences, a major goal of the competition organisers was the development of the youngsters' technological fluency. In other words, participation in SUB12 and SUB14 was expected to contribute to the education of the twenty-first century youngsters in an integrated manner by supporting the development of skills that are needed to respond effectively and creatively to the challenges of an increasingly technological, competitive and demanding world.

During the first edition of the competitions, in 2005, there were some reported difficulties concerning the unavailability of the necessary technological resources. Some participants found it difficult to access a computer in order to submit their answers to the problems through e-mail. In some cases, the lack of a computer or Internet access at home, or at school, prevented the use of e-mail and necessitated the participants' parents sending submissions on paper via post or fax, with the added fear that their children's answers would not arrive on time. A couple of years later, the situation had improved quite significantly, and more recently, the Technological Plan for Education, approved in 2007 by the Portuguese state, helped to provide Internet connection to all the Portuguese schools.

Most recently, national statistics (Statistics Portugal, 2013) have shown that the availability of computer and Internet connection in Portuguese households is reasonably common, being close to 67 % for computer access and 62 % for Internet access. Moreover, the official data also reveal that among the households with children below 15 years of age, the percentage with access to ICT (computer and Internet) has reached close to 90 % (92 and 86 %, respectively, for computer and Internet availability).

The chart in Fig. 2.6 shows that the percentage of respondents to our survey who had a computer, Internet connection and his/her own e-mail is, respectively, about

2.3 The Participants and the Use of Digital Technologies

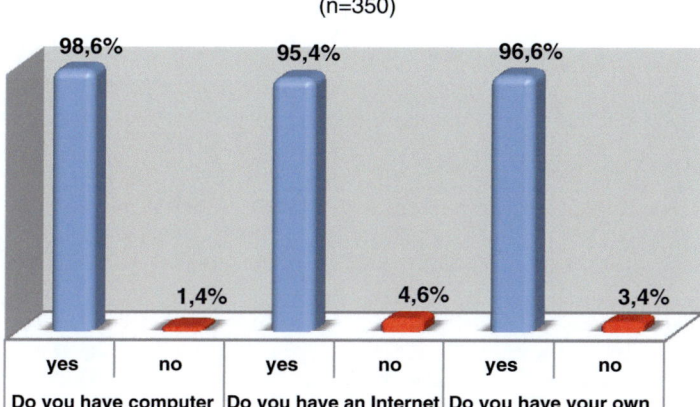

Fig. 2.6 Computer, Internet and e-mail availability at home of the respondents to the survey

99, 95 and 97 %. Thus, only a very small proportion of participants did not have these resources at home. In any case, participants can submit their solutions from a computer at school, or possibly from the home of another colleague or friend, because they are allowed to participate in small groups. We know that some teachers occasionally submitted the answers of their students who did not have a computer or Internet connection. Some parents did the same by sending their children's answers from an existing computer at their workplace. A very small number of participants also used the e-mail of their parents to send their answers.

In 2012, a mathematics teacher of grades 5 and 6, who had followed the SUB12 competition from the very beginning, said in an interview:

> Teacher I: In the early years of the SUB12, many youngsters did not have their own e-mail. Now if we ask about it in a class, we find only one or two youngsters who do not have their own e-mail.

Among the respondents to our survey, we did not find any participants who could not use the Internet (Fig. 2.7). It is precisely when browsing the Internet that these youngsters feel more confident. Some of them appear to do better than adults when Internet searching. In this regard, a teacher explained how she had learnt from her youngsters how to check the tables with the scores of all participants in each problem published on the competition webpage.

> Teacher A: In the class, I project everything on the screen. When the answers come [the teacher refers to the solutions selected by the organization that are published on the website], we all look at them… For example, I did not know I could use the "binoculars" [the Adobe Acrobat search button], so the first time I was there scrolling up and down and saying, "Oh, class 6C, ah, it is not us" (laughs). And the kids told me "Oh, teacher, you skipped it already…" and it would take a lot of time. They explained to me that the binoculars were useful to find the names; now I just search the names of the youngsters from my class and then we see everyone, and it's done!

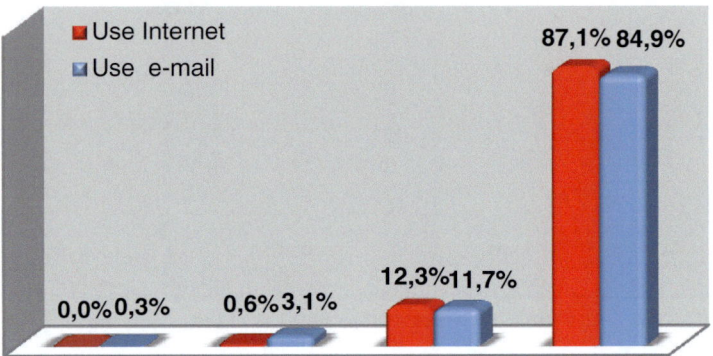

Fig. 2.7 Knowledge of Internet and e-mail use of the respondents to the survey

The fact that these competitions were undertaken through the Internet and required the use of the computer was a motivating factor that could bring some benefits to young people. Rui, a visually impaired student who took part in the two competitions, recognised several benefits in using the computer, especially in his case.

> Rui: The computer helped me because of my difficulties. It allowed me to participate. The use of computers was important because I did not write and had many difficulties to read. I could not write by hand. The computer magnified the statements, which allowed me to read. In terms of writing, it was much easier to do it on the computer. The computer also helped me to discover new resources. I am also indebted to the SUBs for the passion I have for computers.

To this young boy, the computer facilitated his participation. In fact, the option offered by the computer to magnify the characters and images enabled his reading of the problems. Furthermore, the difficulty in writing experienced by Rui was overcome by the possibility of typing and use of other digital tools provided by the computer.

As for Jonas, one of the other interviewees who also took part in the competitions for four successive years, the computer was something almost inaccessible when he was in his 5th grade. The computer belonged to the adults in his home, so his participation in the competitions was a way of getting permission to use it and, at the same time, a challenge for him to learn alone how to make use of it. As he said:

> Jonas: It was hard, it was quite a struggle. I had to review over and over again and to think more… I sent the solutions on the last day, at the last minute, because it took me a week to think about how to create my answer on the computer.

This testimony reveals that for some of the youngsters, participation in the SUB12 and SUB14 online competitions opened the doors to the digital world; in the case of Jonas, in spite of the initial difficulties felt, he reported the satisfaction of having gained access to the computer, which was one of his aspirations as a child at that time.

Having examined the nature of the participation, we turn to the productions of participants while solving the problems with digital tools.

2.4 The Participants' Productions with Digital Technologies

The diversity of solutions submitted to the competitions is remarkable, which made our selection of examples a laborious task. In this section, we present a series of solutions in which technologies were always involved. For this purpose, we created several categories based on the various technological tools commonly used by the participants. These categories emerged from the ways of expressing the solutions to the problems and were grouped in the following way: (1) use of text; (2) use of tables; (3) use of images and diagrams; (4) use of numerical outputs; and (5) use of geometrical constructions. Although these productions have been sent to the competitions by two different means—the online form at the website or the participant's e-mail—this has no bearing on establishing the categories. It is however clear that most of the categories are not entirely exclusive in the sense that several solutions integrate the writing of text with diagrams and images or with tables and other representational elements. To some extent none of the categories, except the use of text, occurs as exclusive. For example, in many of the solutions showing tables, we can also find images or diagrams or symbols. As such, it would be possible for a number of submissions to be included in one or another category.

The sequence of productions presented below is organised from plain text using just paper and pencil to the more technologically sophisticated solutions that are based on the use of specific software tools (e.g. Excel or GeoGebra).

2.4.1 From the Use of Paper and Pencil to Writing with Word and Excel

Participation in the SUB12 and SUB14 competitions was carried out, as we have seen, through written communication. This communication required that participants made a considerable effort to convey all the mathematical reasoning involved in solving the problems that were posed. Mathematical written communication goes far beyond writing numbers, symbols and mathematical operations. Although many studies show that the prevailing view about mathematics and mathematical communication is based on numbers and formulas (Shield & Galbraith, 1998), participants in these competitions frequently also make use of images, diagrams, tables, etc. In fact, there is nowadays a growing demand on youngsters to use forms of mathematical expression that include not only the mathematical symbolism but also verbal sentences in everyday language or other representational forms. International recommendations have stressed the importance of the development of youngsters' competence in the use of symbols and mathematical representations and have emphasised the centrality of argumentation and justification in mathematical tasks (see, e.g. NCTM, 2000; Ntenza, 2006).

The answers submitted to the mathematical problems posed by the SUB12 and SUB14 competitions showed a great variety of ways to communicate and express mathematical thinking. Participants used multiple representations that reflected their reasoning processes and the knowledge they put into action to generate conceptual models and formulate mathematical results. These representations can be seen as inherent to the development of mathematical models that underpin youngsters' approaches to the problems through concrete, verbal, numerical, graphical, contextual, pictorial and symbolic elements, thus depicting aspects of such models. At the same time, these creations are also allowing communicating and presenting mathematical ideas and the explanations for the solutions achieved.

The mathematical competitions SUB12 and SUB14 have contradicted in some way the prevailing view about what it means to write a mathematical solution to a problem or, in other words, about what it means to express a solution mathematically. This stems from the fact that the statement of each problem posed on the competition website is accompanied by a mandatory requirement: "Do not forget to explain your problem-solving process" for the answer to be considered correct and accepted as complete. The explanation of the reasoning, and the presentation of the process undertaken to reach the solution or solutions of the proposed problem, was a crucial requisite. When the answer did not provide a clear explanation, the participant received feedback in which it was stated that even though the answer was correct, it could not be considered as such without the explanation of the whole reasoning developed to achieve it.

The written communication thus became another challenge within the challenge of the competition itself. Each participant was always given complete freedom to present the solution process, as long as it was feasible to send it via e-mail to the organisation.

In researching the SUB12 and SUB14 competitions, we found that the participating youngsters often displayed great creativity in meeting the challenge of solving and cleverly expressing the solutions to the mathematical problems that were posed. The ingenuity of the participating youngsters often surprised the organisation, their teachers and their families.

The ways of presenting the solution and the reasoning process ranged from scanning hand-written work to submissions that were fully supported by technological tools. In the case of solutions made with paper and pencil, participants generally took one of two options; either they sent a digital photograph taken with a mobile phone or digital camera or they attached a PDF file or image file of a scanned page of their work. Using technologies or not for the preparation of the report of the problem-solving process was equally accepted. From the very beginning in 2005, the SUB12 and SUB14 competitions were cast as inclusive competitions. So, among other characteristics, the work made with paper and pencil was accepted on equal terms with productions using technology.

Our first example shows a solution submitted to Problem #8 (Fig. 2.8) of SUB12 from the 2012 to 2013 edition that was totally produced using paper and pencil. As Fig. 2.9 portrays, this participant decided to take a scan of her work and send the image file as an attachment to the e-mail submission.

Mrs. Guida took her four children shopping on a Saturday afternoon and promised them a toy, an ice cream and a drink if they behave well. For each child, find the name, the ice cream, the beverage, and the toy chosen! Use the following information.
1. Ana asked for a caramel ice cream and did not choose an orange juice.
2. Sofia got a set of Dominoes and did not choose an iced-tea.
3. The boy who got a toy car also chose a strawberry ice cream.
4. One of the girls got a doll, one of the boys had chocolate ice cream, one of the boys is named Carlos and one of the girls drank Cola.
5. The boy who got a pack of cards drank a bottle of water and did not choose vanilla ice cream.
6. The child who drank iced-tea did not choose a doll.
7. Leonel did not choose water.

Do not forget to explain your problem solving process!

Fig. 2.8 The statement of the Problem #8 from the SUB12 (2012–2013 edition)

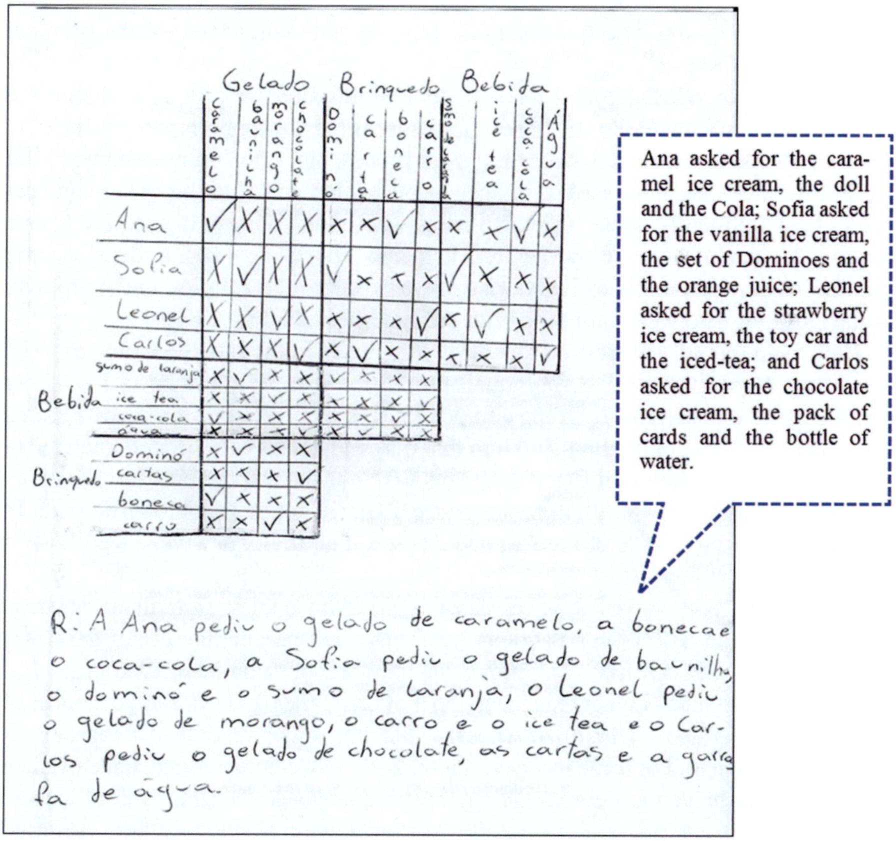

Fig. 2.9 Digital scan of a solution produced with paper and pencil to the Problem #8 of SUB12 (2012–2013 edition)

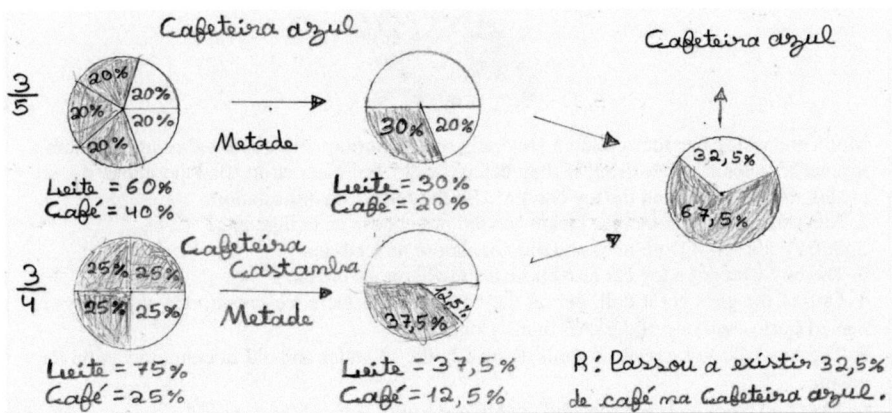

Fig. 2.10 Example of a scanned page with pie charts drawn by hand for Problem #6 of SUB12 (2012–2013 edition)

Other cases of youngsters' work using paper and pencil, sometimes in notebook sheets or on blank sheets or even on a page printed with the problem from the website, tended to share characteristics. As illustrated in Figs. 2.10 and 2.11, some may have calculations, diagrams or other notations and everyday language (see the problem in Fig. 1.11 in Chap. 1).

A scanned or photographed hand-written submission was far from being the most common among the forms of solutions submitted during both competitions. In general, youngsters turned to Word for word processing, often combining text with the insertion of diagrams, tables, images or other elements that they may consider using. Less frequently, there were also cases where the youngsters used Excel as a writing device in which they inserted text and images to present their solving process, as in the example of Fig. 2.12 that was a submitted solution to the problem of the children who went shopping with their mother (see Fig. 2.8).

Data from our survey showed that all participants claimed to know how to write text using Word, with only a very small proportion (<3 %) claiming to have little knowledge of this Microsoft Office component (Fig. 2.13). In terms of use of Excel, our data revealed a contrasting situation that may indicate a very limited knowledge of its features. As shown in Fig. 2.13, about 30 % of the respondents to our survey declared that they knew nothing or very little about Excel as a tool to write with. In the same vein, only 40 % of the respondents said that they knew very well how to use Excel for writing text.

The data from our survey were consistent with what we know from the submission to the SUB12 and SUB14 competitions, year on year, regarding the use of these two components of MS Office. Over time, the number of submitted Word files has consistently been much higher than the number of Excel files. Indeed, a large number of submissions consisted solely of written text within the e-mail rather than being submitted as an attachment. For some received solutions, these were, in essence, quite detailed descriptions of the youngster's reasoning in natural language. For this form of expression, the participants preferred to use the Word processor. Sometimes they also used colours to differentiate or highlight data or any information they considered relevant.

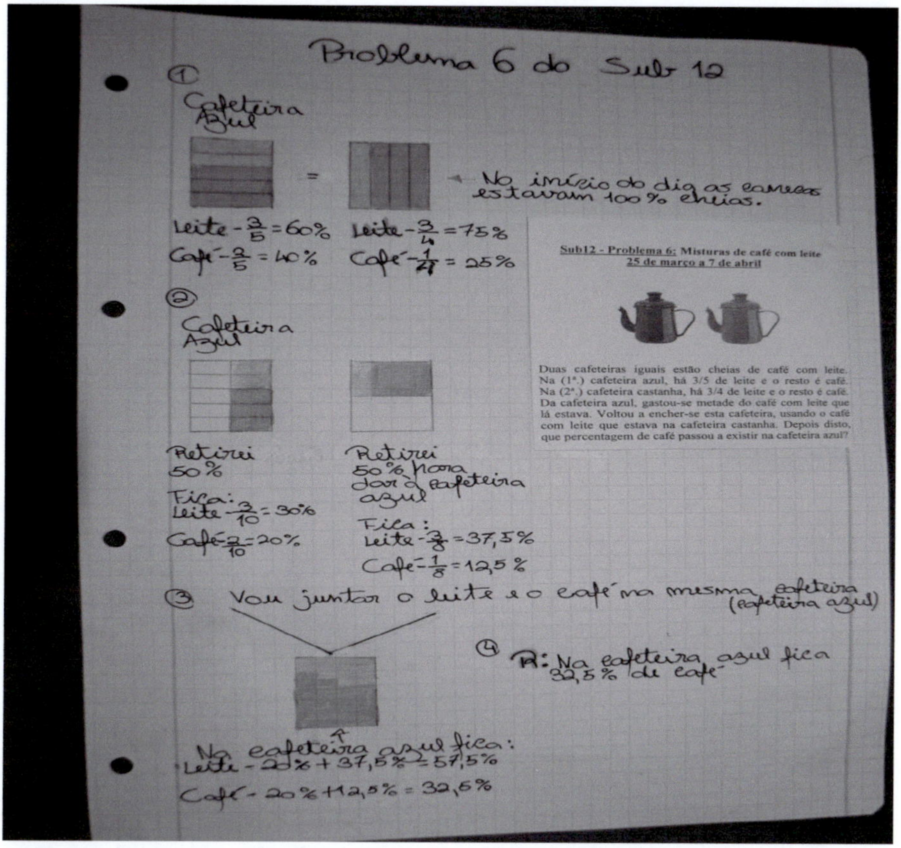

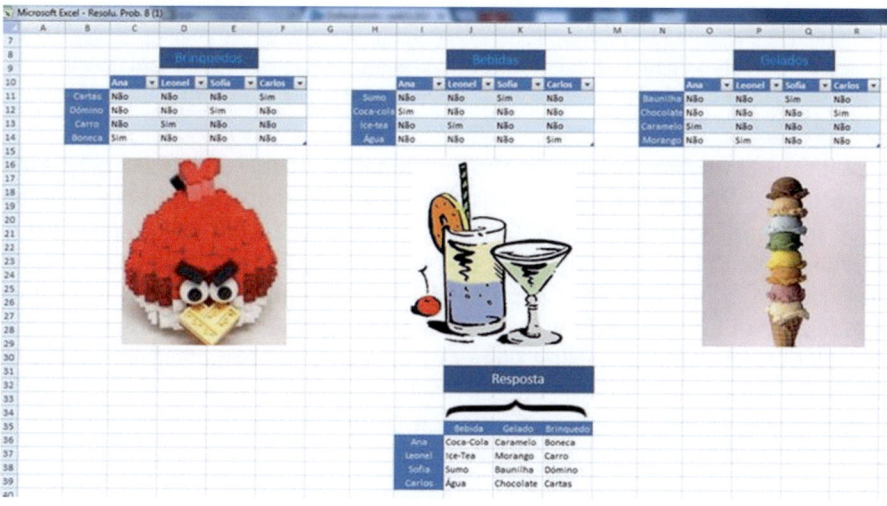

Fig. 2.11 Example of a photographed hand-drawn solution to Problem #6 of SUB12 (2012–2013 edition), illustrating how advantage was taken of squared paper for the diagrams

Fig. 2.12 Example of a solution produced in an Excel sheet where the cells are used as text boxes

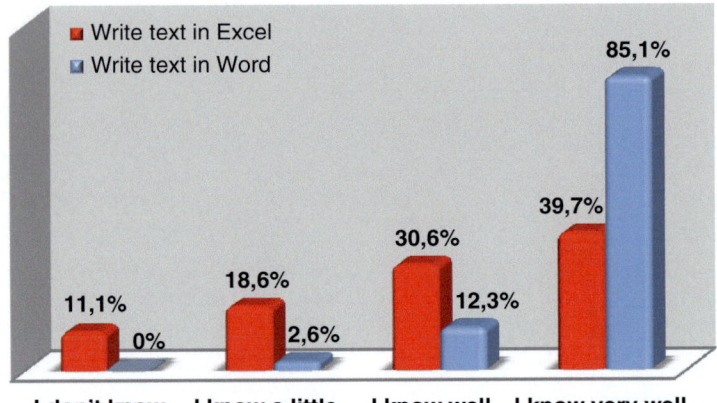

Fig. 2.13 Knowledge of the respondents to the survey on text editing with Word and Excel

Giving detailed descriptions of the youngster's reasoning in natural language was used by a large number of participants in an early phase of learning how to create and edit text with Word. As Mara, one of the participants, confessed, it was during her participation in the online competition that she developed her technical skills in writing with Word. Although she started to write at a quite early age, and began to write the numbers on the computer with her father by the time she was 2 or 3 years old, she was not at all familiar with writing and editing mathematical symbols in Word as she explained:

> Mara: At the beginning, I did it all on that reply box, but later I started doing it in Word. Still it turned out that... my best effort was to solve and explain it in a good written way... (...) There was a phase when I did not know where to find the appropriate symbols in Word. I needed symbols because with just words, I could not explain it, and I had a hard time finding them. I was not familiar with the use of mathematical symbols [in Word] because it had never been necessary before.

With time, we have witnessed the development of these youngsters' digital competences. We found evidence that many of the participants who began in the early stages of SUB12, and continued to participate in the subsequent years, developed their skills in using software and other digital means to express their ideas and solve the mathematical problems. The example in Fig. 2.14 shows how a youngster in the 8th grade was fully capable of producing an explanation for the solution of a problem using plenty of symbolic mathematical language. In fact, the solution showed how this girl could establish a system of linear equations and present all the steps to obtain its solution (see the given problem in Chap. 1, Fig. 1.4).

2.4 The Participants' Productions with Digital Technologies

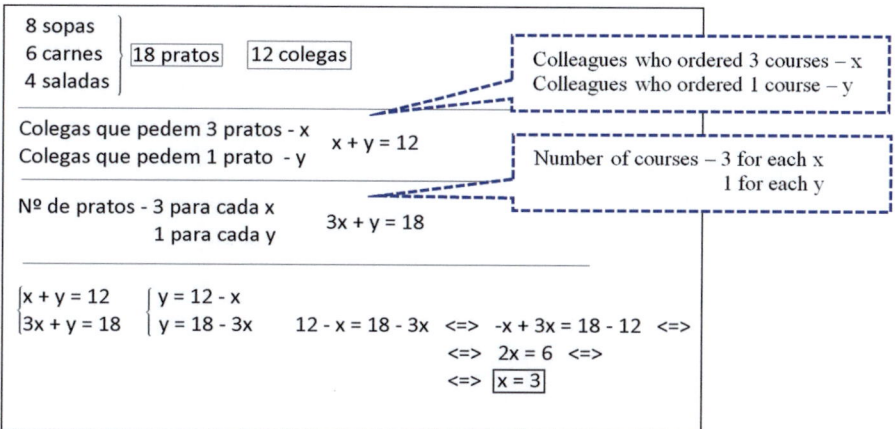

Fig. 2.14 A solution to Problem #7 of SUB14 (2011–2012 edition) submitted as a Word file attached to an e-mail that illustrates the use of symbolic language

2.4.2 The Use of Tables

Creating tables was a fairly common strategy in the solutions submitted to several of the problems proposed in the SUB12 and SUB14 competitions. Over time, the creation and use of tables became more sophisticated and effective as the participants acquired more experience and confidence in their participation. The importance of the use of tables was quite evident in these competitions. In this regard, one of the interviewed participants noted:

> Jonas: In the beginning what I did was describing, I described all the steps. It was a boring thing to be always describing. My father eventually taught me how Excel works and I started to send the tables in Excel to SUB14. Some [problems] had to be actually solved with tables.

Although the problems do not always call for the use of tables, the participants seemed to be quite skilful in their use for a diversity of purposes.

Through the survey, we sought to know how the participants understood their expertise in creating tables in Word and Excel. As shown in Fig. 2.15, there was a noticeable difference between knowing about making tables in Excel and in Word. A large percentage of the participants seemed to be able to create tables in Word, while in Excel the numbers were definitely lower. As Fig. 2.15 illustrates, the percentage of participants who declared to know very well how to create tables in Word was almost twice the percentage of youngsters who say they knew very well how to create tables in Excel. Likewise, a relatively high percentage of participants admitted not knowing or hardly knowing how to create a table with Excel (around 34 %).

The solutions received in the competitions throughout the several editions were in tune with these data. In fact, the solutions considered in the category of "table use" revealed a predominance of tables built with a text editor. In many cases, participants

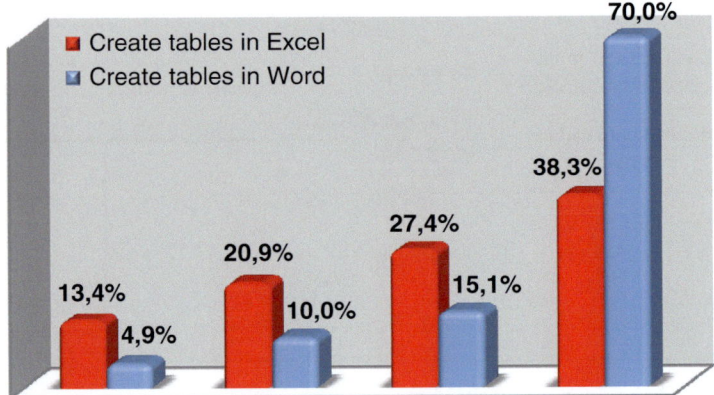

Fig. 2.15 Knowledge of the respondents to the survey on creating tables with Word and Excel

used colours to highlight a particular aspect, including a variety of symbols, characters or images, which became part of the construction of such tables and contributed to the expressive and clear solutions submitted by the youngsters, such as that in Fig. 2.16. Here, the youngster submitted a solution in which a Word table was filled with coloured shapes representing the different dishes in the menus chosen by several friends at lunch (see the given problem in Chap. 1, Fig. 1.4). The role of the Word table was to organise, in a pictorial way, different hypothesis and highlight the correct option. Although the participant also included some text to describe the content of the problem, the pictorial nature of the table clearly depicted how the solution was developed and obtained.

Other solutions using tables combined the use of written language and colours with other elements. These included drawings or pictures that represented the concrete elements referred to in the problem. In Fig. 2.17, for example, the solution has a clear resemblance to the one in Fig. 2.9 that consisted of a table produced by hand, namely, that of signalling the correct and incorrect hypotheses. The distinctive feature of the example made in Word was the presence of significant visual elements that made the answer more noticeable and dramatic (see the given problem in Fig. 2.8).

Although Word emerged as a privileged instrument to construct tables, we found many youngsters, even quite young ones, who used Excel to create tables, especially as a way of organising information, representing a problem in schematic form or systematic counting. Many participants did not make use of the mathematical affordances of the spreadsheet but, instead, took advantage of the grid structure. For some of the problems, tables were a way of devising an underpinning pattern (sometimes a numerical one); having a handy way of creating and visualising a table became an important resource to support the reasoning that led to the solution.

2.4 The Participants' Productions with Digital Technologies

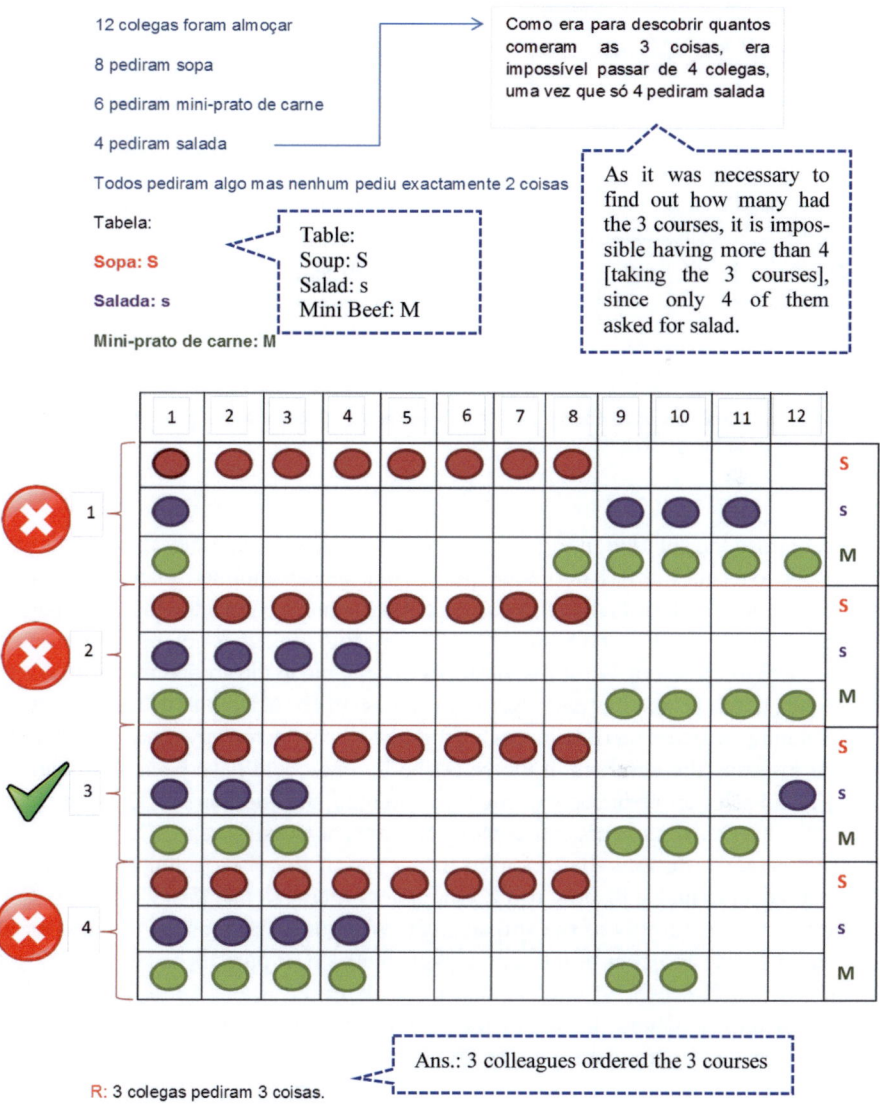

Fig. 2.16 Solution based on the construction of a table in Word using colours and shapes, submitted to Problem #7 of SUB14 (2011–2012 edition)

The Excel table in Fig. 2.18 shows how the organisation of elements in a grid provided a good form of systematic counting in a problem concerning the number of possible combinations of colour, size and style in pairs of jeans.

SUB 12 - PROBLEMA 8

FILHOS	GELADOS				BRINQUEDOS				BEBIDAS			
	Caramelo	Morango	Chocolate	Baunilha	Dominó	Carro	Boneca	Cartas	Sumo de laranja	Coca-cola	Água	Ice-tea
LEONEL	X	V	X	X	X	V	X	X	X	X	X	V
ANA	V	X	X	X	X	X	V	X	X	V	X	X
CARLOS	X	X	V	X	X	X	X	V	X	X	V	X
SOFIA	X	X	X	V	V	X	X	X	V	X	X	X

Fig. 2.17 Solution based on the construction of a table in Word using pictures; submitted to Problem #8 from the SUB12 (2012–2013 edition)

In contrast, the solution in Fig. 2.19 concerns the determination of the first number belonging to a later line of a numerical pattern that was exemplified solely by its initial sequence. Using an Excel table, this participant laboriously completed the entire pattern line by line until reaching the desired line. Clearly, this strategy is not the most efficient one to solve the problem, but it was perfectly acceptable to the competition organisers. Moreover, it tells us that this participant saw the use of Excel as suitable for the purpose of generating the entire set of numbers belonging to the numerical pyramid. The lengthy task of inserting the numbers in the cells one by one could have been simplified by using Excel to generate linear sequences by dragging the "fill handle" to extend automatically a series of numbers, either along columns or along rows.

Unexpected solutions that resorted to the use of Excel have led us to look for a better understanding of how these youngsters learnt to use Excel and how they perceived the usefulness of the spreadsheet. We found that most of the interviewees learnt to use Excel at home, either with the help of parents or by exploring it on their own. However, it is clear that they identified it as a school tool mainly useful for their teachers. In one of the interviews with a 10-year-old participant, we realised some interesting facts about how he created his solutions on the computer at home.

>Interviewer: Why do you use Excel?
>Raúl: I think it's easier [to solve and present the problem in Excel]. The last problem, I did it using Word—but most of the time I have to do tables and it is easier with Excel.
>Interviewer: And what else can you do with Excel?
>Raúl: I do not know much, I know that teachers use it to make their grading grids and to make other tables.
>Interviewer: How do you know that teachers use Excel to make grids?
>Raúl: It was my mother who showed me.

2.4 The Participants' Productions with Digital Technologies

		Azul	Azul pré lavado	Preto	preto pré lavado	TOTAL
	tamanho 1	1	1	1	1	4
	tamanho 2	1	1	1	1	4
modelo 1	tamanho 3	1	1	1	1	4
	tamanho 4	1	1	1	1	4
	tamanho 5	1	1	1	1	4
	tamanho 1	1	1	1	1	4
	tamanho 2	1	1	1	1	4
modelo 2	tamanho 3	1	1	1	1	4
	tamanho 4	1	1	1	1	4
	tamanho 5	1	1	1	1	4
	tamanho 1	1	1	1	1	4
	tamanho 2	1	1	1	1	4
modelo 3	tamanho 3	1	1	1	1	4
	tamanho 4	1	1	1	1	4
	tamanho 5	1	1	1	1	4
	tamanho 1	1	1	1	1	4
	tamanho 2	1	1	1	1	4
modelo4	tamanho 3	1	1	1	1	4
	tamanho 4	1	1	1	1	4
	tamanho 5	1	1	1	1	4
					TOTAL	80

Fig. 2.18 Solution based on the construction of a table in Excel to count the number of combinations in Problem #7 from the SUB12 (2011–2012 edition)

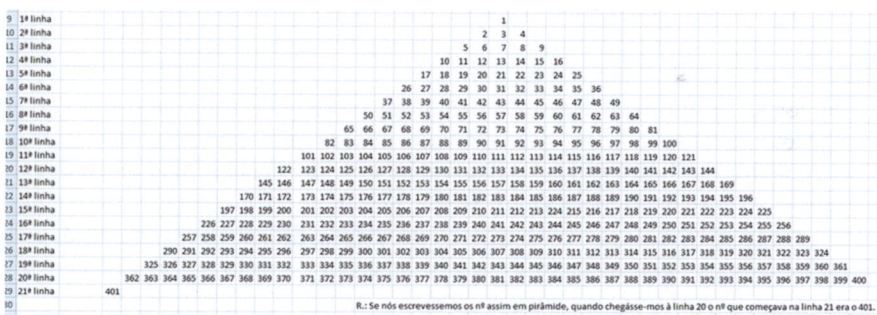

Fig. 2.19 Solution based on the construction of a table in Excel to present a sequence of numbers in Problem #8 from the SUB12 (2011–2012 edition)

2.4.3 The Use of Images and Diagrams

There was plentiful use of images in the solutions presented to the mathematical problems of the SUB12 and SUB14 competitions, including many cases where participants resorted to MS Paint. The graph in Fig. 2.20 illustrates the participants' answers to the survey questionnaire in relation to their knowledge about how to insert images and the use of Paint. As can be seen, inserting images is a competence that participants

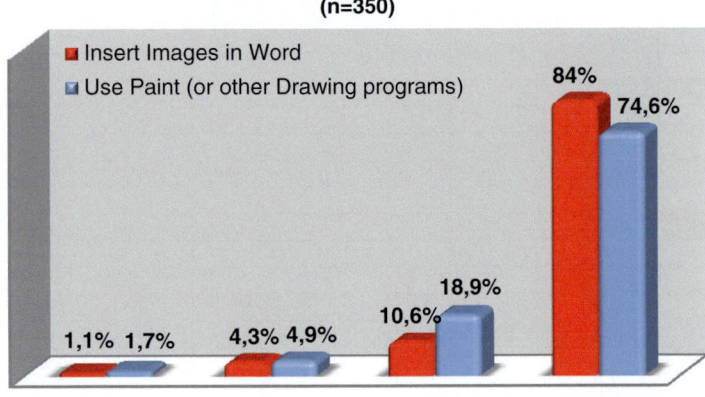

Fig. 2.20 Knowledge of the respondents to the survey on using images with Word and Paint

considered having in full. In fact, the use of images to convey ideas and reasoning was widely applied in the solutions received within files in Word and even Excel.

The participants showed their creativity in the expression of their solutions, through using images in interesting ways. This was perhaps one of the aspects that most clearly showed a move beyond regular problem-solving routines and procedures in the school classroom; the use of images is a natural consequence of Internet access that enables the search for pictures, signs and illustrations that convey ideas and objects that are relevant and useful in solving the problem and expressing ideas about the ways of obtaining the answer.

Our evidence illustrates ways in which many of the youngsters who engaged in SUB12 and SUB14 for a time span of 4 years underwent a remarkable evolution in their technological skills. Rui is such an example of the phenomenon of starting to use images to present his ideas and reasoning in solving the problems of the competitions. Noticeably, Rui was a youngster with writing difficulties as the result of a visual impairment who reported to us that he acquired many of his digital skills during his participation in the competitions.

In Problem #1 of SUB14 posed during the 2011–2012 school year, Rui used a diagram reproduced in Fig. 2.21 competition (see Chap. 7 for an in-depth analysis of this motion problem). In his solution, Rui used MS Paint to express his reasoning by including two different images of faces to represent the two friends walking towards each other, as mentioned in the problem.

Another participant in the SUB12 competition, in answering Problem #6 of the 2012–2013 edition that involved the combination of blends of coffee and milk in different proportions (see the problem shown in Chap. 1, Fig. 1.11), sent an attachment in Word that was composed of several images together with text boxes displaying the relevant results and arrow shapes to describe the blending of different mixtures and finally presenting the solution (see Fig. 2.22).

2.4 The Participants' Productions with Digital Technologies

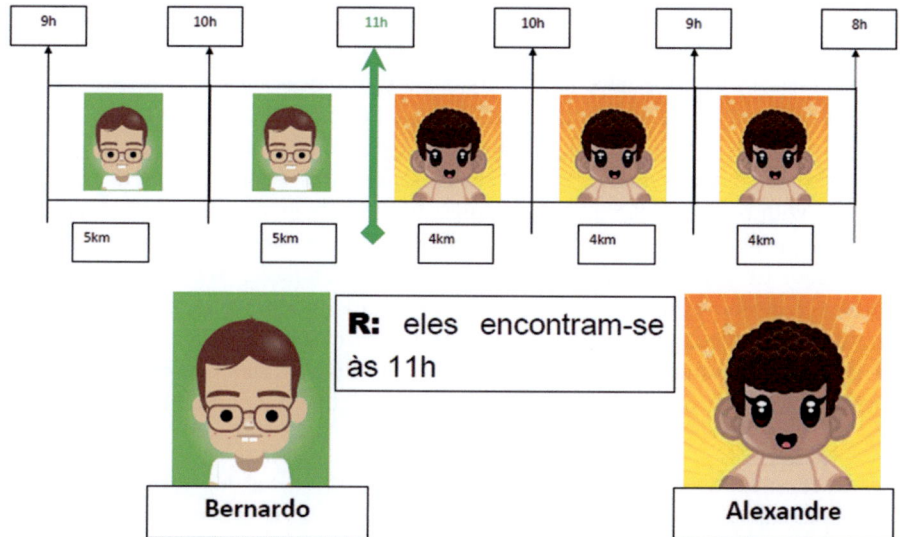

Fig. 2.21 A solution using pictures to present a diagram with Rui's approach to a motion problem

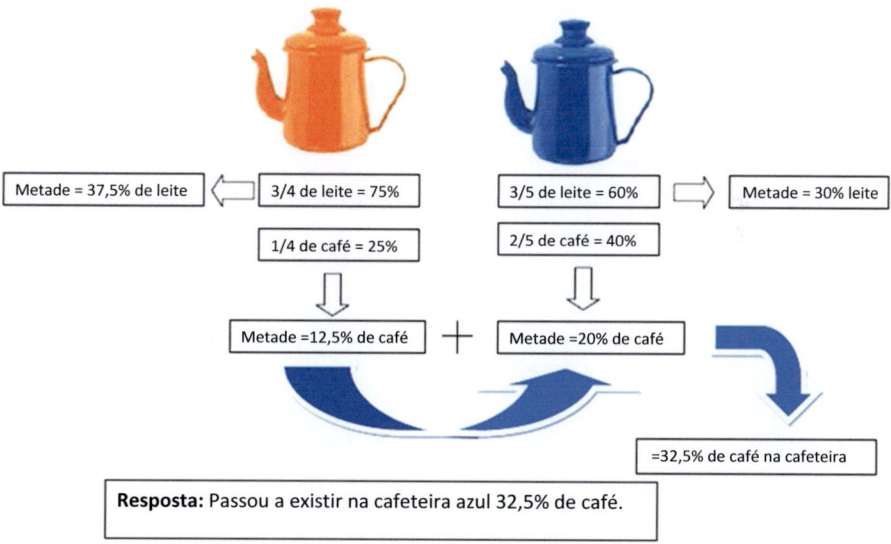

Fig. 2.22 A solution using a diagrammatic description of the reasoning in Problem #6 from SUB12 (edition 2012–2013)

The strength of visual expression in the collected solutions was so substantial that we even found the use of Excel for creating diagrams rather than for tables or for calculations. In fact, we have seen that situation in cases where Excel was used to express ratios in iconic and rather creative ways. For example, filling a tank with two taps, A and B, having distinct flow rates (see the given problem in Chap. 1, Fig. 1.12) was translated figuratively by coloured cells from an Excel sheet, as shown in Fig. 2.23.

Also with regard to the use of schemes, images, diagrams and other visual and iconic representations, we observed that PowerPoint was a commonly chosen instrument by many participants to deliver their solutions. When analysing the answers received over several editions of the competitions, we recurrently found a number of solutions presented as PowerPoint files. In general, there was a group of participants who frequently used this resource to display and send their solutions. The questionnaire data presented in Fig. 2.24 show that the participants claimed to

Fig. 2.23 A solution using a visual representation of ratios in Problem #3 from SUB14 (edition 2011–2012)

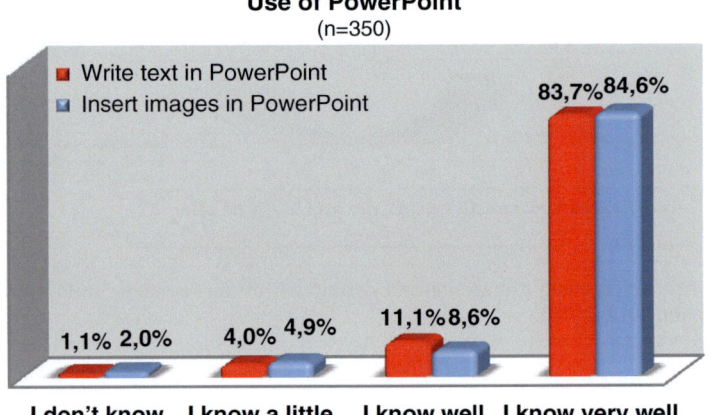

Fig. 2.24 Knowledge of the respondents to the survey on using text and images in PowerPoint

2.4 The Participants' Productions with Digital Technologies

know very well how to use PowerPoint. Additionally, the number of participants who did not know or barely knew this tool was residual (around 6 %), both in terms of creating text in a slide or inserting images to illustrate ideas and explanations.

Figures 2.25a–c show an example of a solution submitted by a participant who often sent her solutions as PowerPoint files. Her answer to a problem related to filling a tank with two different taps was presented as a sequence of three slides (see the given problem in Chap. 1, Fig. 1.12). The first two slides situated the context of the problem and identified the relevant given data. The third slide described the solution process based around a schematic picture where the combination of the two taps was accurately explained and represented.

This category of solutions, embedded in a visual and pictorial form of solving and expressing, seems to be persistent in all the editions and across all ages of participants. Accompanying this there were particular cases of youngsters who showed a clear progression and developing perfection in the presentation of their solutions in visual and diagrammatic ways. Such solutions encompassed a great diversity of skills and many different technological tools, not all of them obvious for the production of pictorial representations.

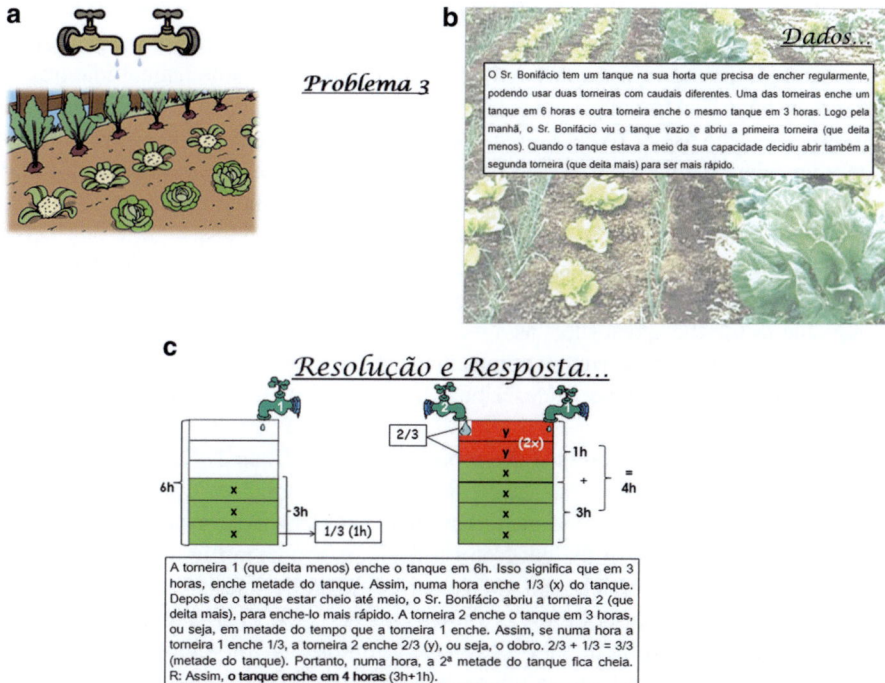

Fig. 2.25 (a) First slide of a PowerPoint file in a solution to Problem #3 from SUB14 (edition 2011–2012). (b) Second slide of a PowerPoint file in a solution to Problem #3 from SUB14 (edition 2011–2012). (c) Third slide of a PowerPoint file in a solution to Problem #3 from SUB14 (edition 2011–2012)

The solutions reproduced here are a sample from a range of very different problems: one on motion, a problem on ratios and a problem about combined flow rates. Across this variety, young participants found several ways to translate their thinking into visual forms and schemes that they built with various digital tools. This showed a significant willingness to take advantage of digital technologies to solve and express the solutions to the problems posed during the SUB12 and SUB14 competitions through the use of images, diagrams and pictorial representations.

2.4.4 The Use of Numerical Software

The participants' use of a spreadsheet shows a certain progress over the years, since the beginning of the SUB12 and SUB14 competitions in 2005. This should be understood at two levels. On the one hand, we have witnessed the growth in the number of participants who use Excel in some way to present their solutions. On the other hand, we should also point to how such use has been developing. Not only did the number of Excel files increase, but the type of affordances used also evolved. Nevertheless, as mentioned previously, there remained a large percentage of Excel users that only took advantage of the possibility of quickly creating a table with text entries.

Increasingly, however, we found more use of Excel to construct numerical relations through the use of formulas and other numerical and mathematical features of the spreadsheet. The use of graphics (e.g. pie charts) also emerged in some solutions. In most cases, the use of Excel as a mathematical tool for calculating and modelling algebraic relationships appeared to be supported by teachers, particularly in cases where the young participants were encouraged to work on the proposed problems in the mathematics classes or in other school settings.

The data from our survey that is displayed in Figs. 2.26 and 2.27 show that Excel was not fully familiar to many of the young participants, despite a moderate percentage being able to use the spreadsheet as a means to carry out numerical calculations; yet this percentage decreases when it comes to using the spreadsheet more productively, including creating formulas to describe relations and analyse variable values that are dependent on others.

An example of a solution (see the problem given in Chap.1, Fig. 1.3) that takes advantage of the Excel functions based on the use of formulas and variable columns and wherein an algebraic solution to the problem is obtained by inspecting the values generated by replication of the generator formula is shown in Fig. 2.28.

Another resolution which also makes use of formulas and is based on the construction of the variable-columns is shown in Fig. 2.29. This solution refers to a classical algebraic problem that establishes conditions for relating the ages of individuals in the present and some years later.

The use of Excel in problems involving quantity variation has proved very significant to our project and has led us to pay special attention to this powerful digital resource in problem-solving and expressing with technologies. Chapter 6 is devoted entirely to the study of a variety of solutions collected in the SUB12 and SUB14 competitions and seeks to give a perspective on how this digital tool supports the

2.4 The Participants' Productions with Digital Technologies

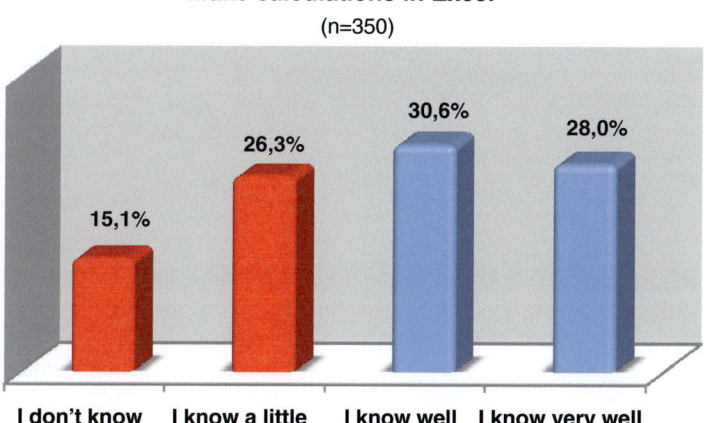

Fig. 2.26 Knowledge of the respondents to the survey on using Excel to make calculations

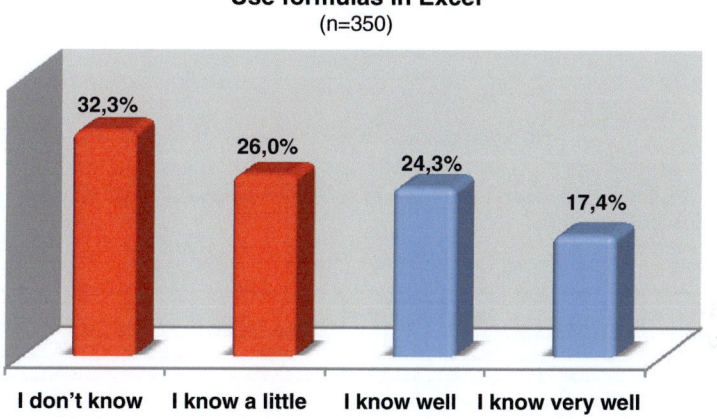

Fig. 2.27 Knowledge of the respondents to the survey on using Excel to generate variable values through the use of formulas

youngsters' conceptual models and how it can be interpreted in light of the co-action between the subject and the tool in problem-solving.

2.4.5 The Use of Geometrical Software

Currently, GeoGebra, a form of dynamic geometry software that allows the combining of geometry and algebra, is fairly used in mathematics classrooms. The advantages and potentialities of dynamic geometry software in the teaching and learning of mathematics are well recognised and have received specific attention from

V1	V2	V3	V4	F1	F2	F3	F4	Soma F
2	4	6	8	12	18	16	14	60
4	6	8	10	18	24	22	20	84
6	8	10	12	24	30	28	26	108
8	10	12	14	30	36	34	32	132
10	12	14	16	36	42	40	38	156
12	14	16	18	42	48	46	44	180
14	16	18	20	48	[illegible]	52	50	204
32	34	36	38	102	108	106	104	420
34	36	38	40	108	114	112	110	444
36	38	40	42	114	120	118	116	468
38	40	42	44	120	126	124	122	492
40	42	44	46	126	132	130	128	516

V1	V2	V3	V4	F1
2	=E6 + 2	=F6 + 2	=G6 + 2	=SOMA(E6+F6+G6)
=E6+2	=E7 + 2	=F7 + 2	=G7 + 2	=SOMA(E7+F7+G7)
=E7+2	=E8 + 2	=F8 + 2	=G8 + 2	=SOMA(E8+F8+G8)
=E8+2	=E9 + 2	=F9 + 2	=G9 + 2	=SOMA(E9+F9+G9)

Legenda:
V1 - número colocado no vértice 1
V2 - número colocado no vértice 2
V3 - número colocado no vértice 3
V4 - número colocado no vértice 4
F1 - número atribuído à face 1
F2 - número atribuído à face 2
F3 - número atribuído à face 3
F4 - número atribuído à face 4
Soma F - soma dos números atribuídos às faces

RESPOSTA: Nos vértices do tetraedro foram colocados os números: 8, 10, 12 e 14.

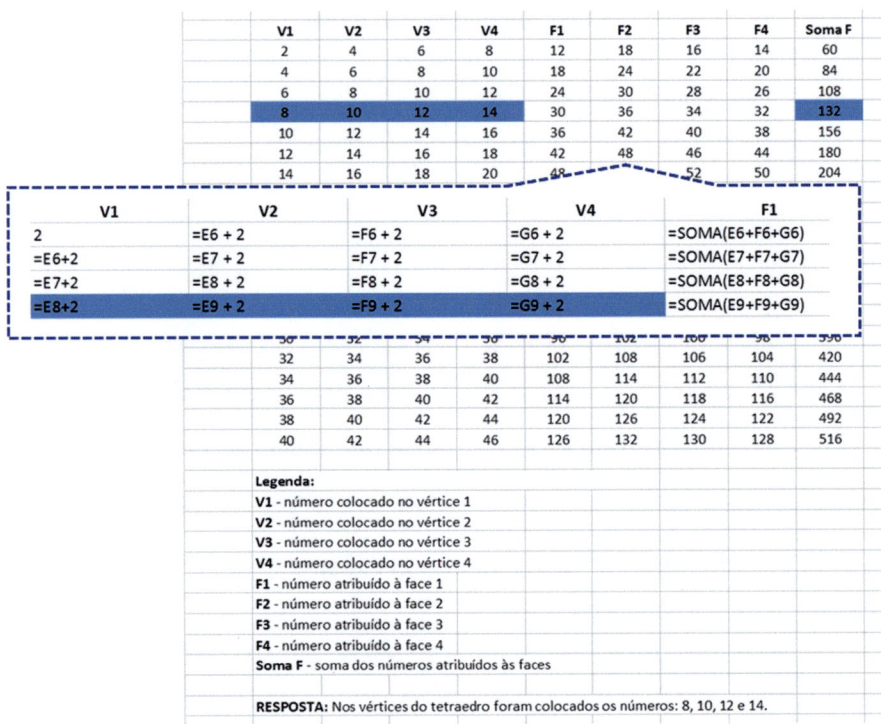

Fig. 2.28 A solution with Excel to Problem #3 of SUB14 (edition 2010–2011) involving several variables that are inter-related

idades			idades + 9			
Soraia	Irmão	produto das idades	Soraia	Irmão	produto das idades	diferença do produto das idades
1	4	4	10	13	130	126
2	5	10	11	14	154	144
3	6	18	12	15	180	162
4	7	28	13	16	208	180
5	8	40	14	17	238	198
6	9	54	15	18	270	216
7	10	70	16	19	304	234
8	11	88	17	20	340	252
9	12	108	18	21	378	270
10	13	130	19	22	418	288
11	14	[illegible]	[illegible]	[illegible]	[illegible]	[illegible]06
12	15	[illegible]	[illegible]	[illegible]	[illegible]	[illegible]24
13	16	[illegible]	[illegible]	[illegible]	[illegible]	[illegible]42

idades + 9			
Soraia	Irmão	produto das idades	diferença do produto das idades
=A3+9	=B3+9	=D3*E3	=F3-C3
=A4+9	=B4+9	=D4*E4	=F4-C4
=A5+9	=B5+9	=D5*E5	=F5-C5
=A6+9	=B6+9	=D6*E6	=F6-C6
=A7+9	=B7+9	=D7*E7	=F7-C7
=A8+9	=B8+9	=D8*E8	=F8-C8
=A9+9	=B9+9	=D9*E9	=F9-C9
=A10+9	=B10+9	=D10*E10	=F10-C10
=A11+9	=B11+9	=D11*E11	=F11-C11
=A12+9	=B12+9	=D12*E12	=F12-C12
=A13+9	=B13+9	=D13*E13	=F13-C13

Fig. 2.29 A solution with Excel to another algebraic problem involving unknown quantities

2.4 The Participants' Productions with Digital Technologies

researchers in mathematics education (Baccaglini-Frank & Mariotti, 2010; Iranzo & Fortuny, 2011; Jones, 2011; Jones, Mackrell, & Stevenson 2009). Accordingly, and given that there were a limited number of solutions using this software throughout the SUB12 and SUB14 competitions, we tried to find out whether the respondents know and actually use GeoGebra. Answers to our survey show a high percentage (around 59 %) of participants that do not know or know only a little about this software (Fig. 2.30). Of the 41 % who were familiar with GeoGebra, 20 % say that they know it well and around 21 % that they know it very well. The answers reporting a good or very good knowledge of this tool obviously are not a guarantee that those participants in fact use GeoGebra in their mathematics daily tasks; the numbers tell us little about the significance of the knowledge claimed in the youngsters' answers. However, the solutions that were received in the SUB12 and SUB14 competitions with this digital tool mirror a certain fluency that is mostly related to an awareness of the tool's usefulness for constructing rigorous geometrical figures, to manipulate them, and use the dragging mode. Moreover, some of the participants were also conscious of the possibility of creating $x-y$ graphs by means of plotting ordered pairs inserted in a table.

The solutions making use of GeoGebra show that the number of young people who employed this software to produce and communicate their mathematical thinking was small. GeoGebra was used by the participants to solve mostly geometry problems and one or two others where the graphing of functions was useful. In a variety of cases, GeoGebra was simply used as a drawing tool for creating simple figures (that could also be produced with other tools for drawing in many different software packages). Such use can occur, for example, for the drawing of a schematic representation of fractions by marking areas in a rectangle with various colours. Another similar situation was the construction of a diagram that represents a linear displacement and shows plotted points in a straight line more accurately than in a rough sketch, as illustrated in Fig. 2.31 (see the given problem in Chap. 7, Fig. 7.1).

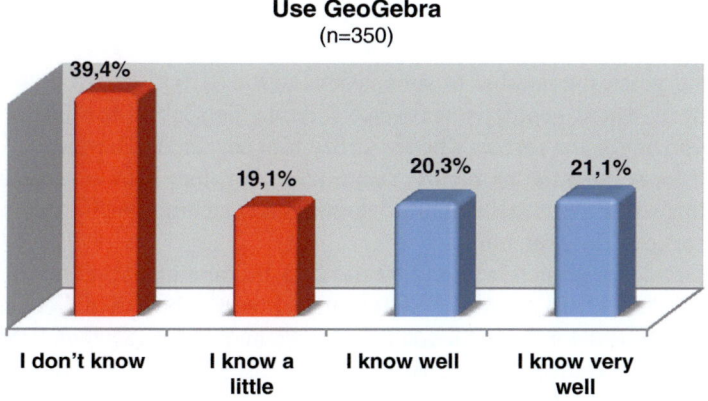

Fig. 2.30 Knowledge of the respondents to the survey on generally using GeoGebra

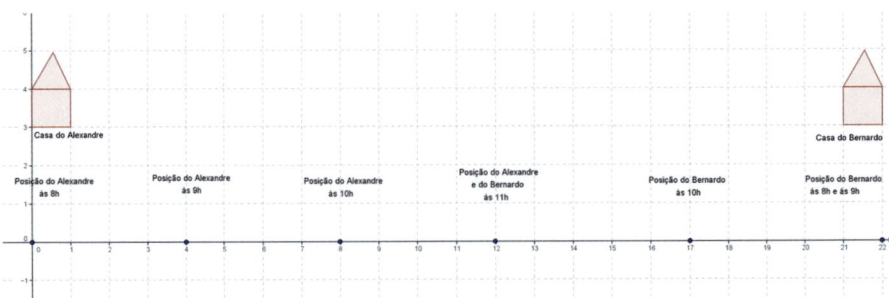

Fig. 2.31 A schematic representation of the solution to a motion problem where GeoGebra is a tool to make a sketch of the different positions along a straight line

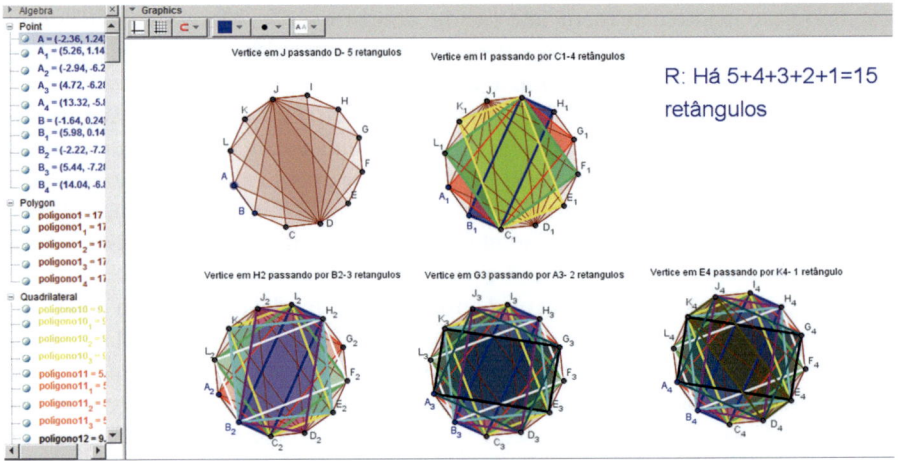

Fig. 2.32 A solution with GeoGebra to Problem #3 of SUB12 (edition 2012–2013) showing the drawing of several types of rectangles that are being overlaid

Over the years, the number of submissions to the SUB12 and SUB14 competitions using dynamic geometry software, such as GeoGebra, has been relatively small. Surprisingly, the solvers who chose this tool appeared to have some expertise in using it for solving the problems. This finding, together with the current importance of this software in mathematics learning and teaching, justifies our attention and deserves our close attention.

The examples presented in Figs. 2.32 and 2.33 are submitted solutions to Problem #3 from the 2012–2013 edition of SUB12 that illustrate the use of GeoGebra. The problem involved the systematic counting of the number of rectangles that can be formed using the vertices of a 12-sided regular polygon (see the given problem in Chap. 1, Fig. 1.2). This is a problem in which the diversity of representations used to present the reasoning, and express the solution obtained, was very evident.

2.4 The Participants' Productions with Digital Technologies

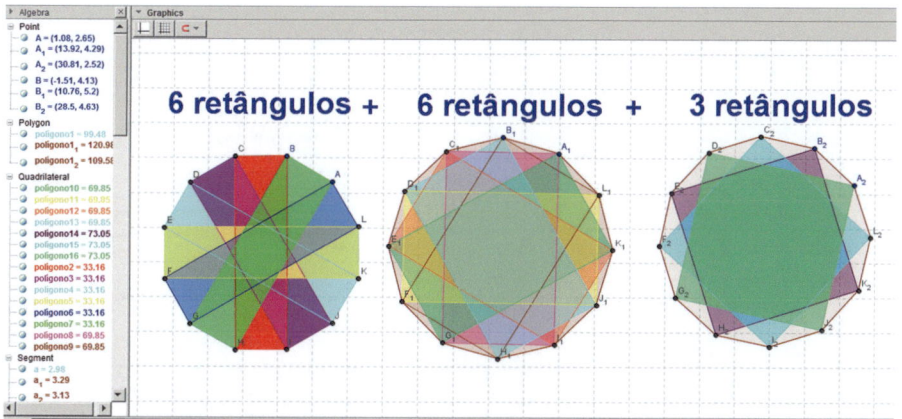

Fig. 2.33 A solution with GeoGebra to Problem #3 of SUB12 (edition 2012–2013) showing another way of identifying the several types of rectangles

GeoGebra was one of the tools used by a number of youngsters with great effectiveness, which again reinforces our conviction that youngsters' participation in such activities enriches and stimulates the simultaneous development of mathematical and technological skills.

Several participants chose to trace all possible rectangles, thus facing some difficulties in maintaining the visibility of individual rectangles due to the large number of lines in the figure. Some of the youngsters who picked GeoGebra to solve the problem realised they could do this overlapping of rectangles using a sequential process in which they introduced new rectangles to the previous figure. Other youngsters preferred to identify the various possible types of rectangles according to the lengths of the sides, recognising the existence of three different types and thus lessening the tangle of lines in the figures.

GeoGebra was also useful to some of the participants in other problems. Both in SUB12 and SUB14, some participants used GeoGebra to tackle geometry problems that required determining a specific measure (e.g. an angle or a length).

For instance, Problem #5 of SUB12 (edition 2012–2013) showed a square with an inscribed equilateral triangle and was asking for the size of a particular angle. Some participants decided to construct the figure using GeoGebra. The answer in Fig. 2.34, submitted by a 6th grader, shows the detailed description of the whole process from the construction of the square and the triangle to the moment where he found the size of the angle that was required.

Other uses of GeoGebra have also appeared in the SUB14 competition. Figure 2.35 shows the solution to another geometry problem involving areas and lengths. Looking at the construction protocol of some of these GeoGebra files opens up a very useful way to get an idea of how these youngsters are able to tackle problems in ingenious and original ways by being aware of the software's affordances. For instance, this solution was obtained by means of the construction of the

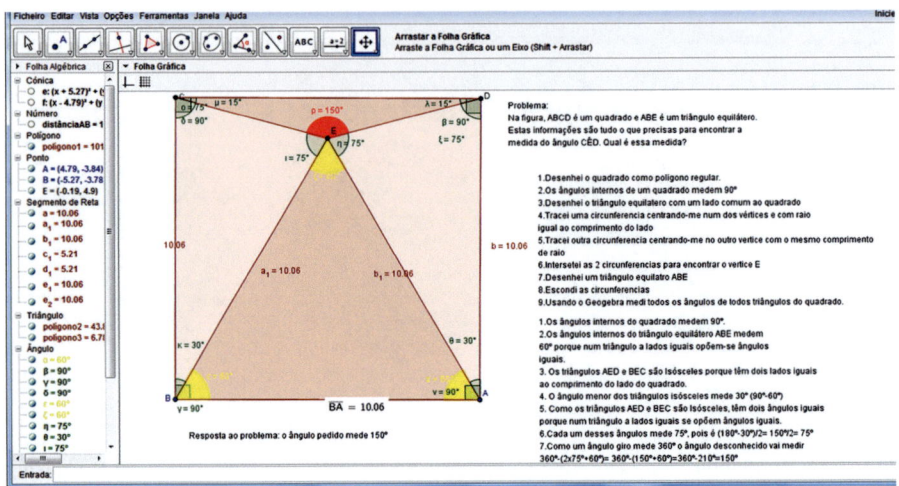

Fig. 2.34 A solution with GeoGebra to Problem #5 of SUB12 (edition 2012–2013) showing the construction together with a detailed explanation of the reasoning

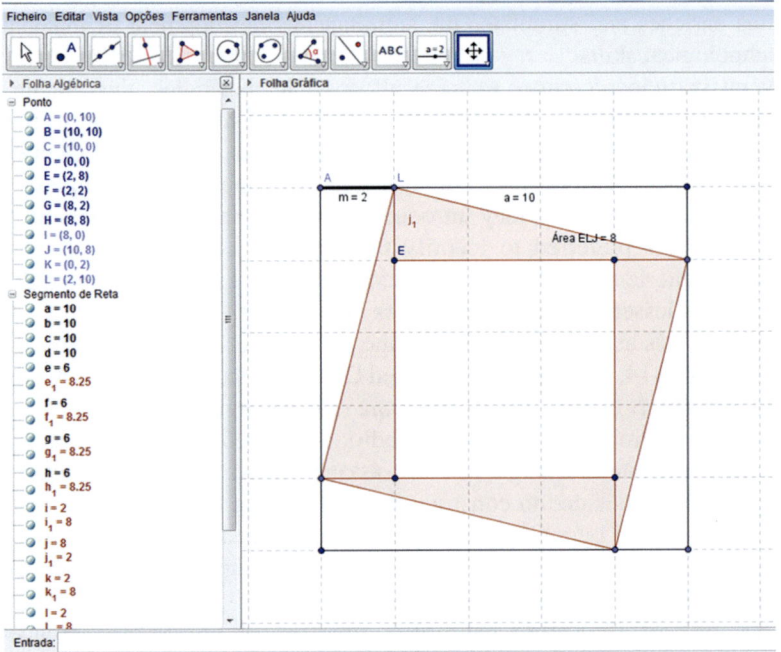

Fig. 2.35 A solution with GeoGebra to Problem #5 of SUB14 (edition 2012–2013) showing the construction used to find the requested length

figure: firstly the larger square, then the inner square, followed by the four triangles and some measurements (area of the smaller square and lengths of segments). They were all constructed by placing points on the grid provided by GeoGebra's graphic view, which means that despite not being a "robust construction," it supported the finding of a solution to the problem.

In fact, when asked about how they learnt about GeoGebra, some participants claim to have discovered it at school but became enthusiasts by downloading and exploring its potential at home, on their own, while solving the competition problems. Furthermore, they were well aware of the usefulness of this software in the devising of a strategy for obtaining the solution to a particular problem and combining its use with that of other digital tools. As Jessica explained:

Jessica:	Hum… usually I look for the notepad and a pen, then [go to] Word and then I always… well I always use GeoGebra or some other software to add something to the text, for presenting a more complete work.
Interviewer:	So… you use it [GeoGebra] only after you solved the problem?
Jessica:	Yes, but… it depends. If GeoGebra or some other tools would help me understand the problem, then I'd use it firstly and later I'd move to Word.
Interviewer:	Ok, so you also use it while you're still looking for the solution…
Jessica:	Yes, for instance, in this case [points at a particular problem] I started by going to GeoGebra to understand it properly, and then I discovered "Oh, that is a triangle right there, therefore I have to subtract the area of that triangle." In that case, I started with GeoGebra for a better understanding.

The availability and usability of this dynamic geometry software seem to be promoting powerful mathematical approaches to the competitions' problems that refer to geometrical notions. Chapter 5 presents an analysis in greater depth of the solutions of the participants to a geometry problem using GeoGebra, where the participants not only show different levels of expertise in using the software but also how such knowledge blends with the development of a conceptual model underlying the solution.

In general, all the interviewed youngsters recognised the importance of their participation in the SUB12 or SUB14 competitions in the development of their technological skills as much as, or possibly more than, their progress in solving mathematics problems.

2.5 Concluding Comments

In this chapter, we presented the profile of the youngsters who participate in these online competitions, by combining data stemming from a survey that reports their levels of expertise in using certain technological tools with selected productions that highlight the quantitative results.

Overall, our results show that these youngsters are far from being all alike or to have the same preferences and experiences in the digital world. In fact, many of these participants feel quite comfortable with solving the problems by hand writing, using paper and pencil, making their sketches or using more formal mathematical processes with conventional means. Then they photograph, digitalise or send scanned pictures of their work. Wonderful solutions have been presented in this style during the competitions. But there are many others for whom the possibility of using the computer is a challenge, an opportunity and even the fulfilment of the desire to use it. We are conscious that this is part of the attraction of the competitions to many of the participants.

There is strong evidence that these youngsters are quite familiar with the most commonly used technologies, including text editors, image editors and presentation editors. We note lesser ease with the spreadsheet and specific programs for mathematical work, such as GeoGebra. However, there were also several examples that reported the use of these technical tools to devise very interesting solutions, namely, from the younger participants.

The combination of these different types of data provides a strong indicator of the role of digital technologies both in solving these mathematical problems and in expressing the mathematical thinking developed during that process. Furthermore, and in line with previous research, the development of technological skills seems associated with the representation, innovation and creativity, thus not limited to a trivial use of technology (Jacinto & Carreira, 2008). The mathematical problem-solving developed within SUB12 and SUB14 is an activity that puts the expression of thinking at its centre, making it difficult to separate the solving phase from the reporting stage. The solutions presented earlier in this chapter depict these two closely linked aspects of mathematical problem-solving, which become more relevant and visible when digital tools are available to support such processes.

When we argue that the expression of thinking is an integral part of mathematical problem-solving, we are pushed to understand how such an expression takes place today, in digital environments, as expressing mathematical thinking is inseparable from the means to achieve it.

In later chapters, with some selected problems, we examine in detail the use of various digital tools chosen by the participants in the two online competitions. Chapters 5, 6 and 7 focus on the students' ways of using digital tools around the mathematics of invariance, variation and co-variation during the joint process of solving a problem and expressing the strategy and mathematical thinking being developed at the same time.

Throughout the various editions of the two online competitions, many mathematics teachers who supported their students during this time (including using the problems in their classes) have attributed great value not only to the engaging nature of the problems but also to the opportunities to develop the mathematical communication and expository discourse of their students. For many of the teachers, the fact that youngsters were describing the process of achieving the solution to a problem was seen not only as one of the challenging aspects of the competitions but also a feature that provided an opportunity for them to innovate their teaching in terms of exploring different mathematical representation systems and taking advantage of technological affordances to convey mathematical meaning through problem-solving.

In the next chapter, we address the perceptions of teachers and their practical support and encouragement to students, showing how the competitions SUB12 and

SUB14 represent, in many ways, useful educational resources that helped to bolster the bond between the youngsters and their social network of family, friends and teachers.

References

Baccaglini-Frank, A., & Mariotti, M. A. (2010). Generating conjectures in dynamic geometry: The maintaining dragging model. *International Journal of Computers for Mathematical Learning, 15*, 225–253. http://dx.doi.org/10.1007/s10758-010-9169-3.

Iranzo, N., & Fortuny, J. (2011). Influence of GeoGebra on problem solving strategies. In L. Bu & R. Schoen (Eds.), *Model-centered learning: Pathways to mathematical understanding using GeoGebra* (pp. 91–104). Rotterdam: Sense Publishers.

Jacinto, H., & Carreira, S. (2008). "Assunto: resposta ao problema do Sub14": A Internet e a resolução de problemas em torno da competência matemática dos jovens. In A. P. Canavarro, D. Moreira, & M. I. Rocha (Eds.), *Tecnologias e educação matemática* (pp. 434–446). Lisboa, Portugal: Secção de Educação Matemática da SPCE.

Jones, K. (2011). The value of learning geometry with ICT: lessons from innovative educational research. In A. Oldknow & C. Knights (Eds.), *Mathematics education with digital technology* (pp. 39–45). London: Continuum.

Jones, K., Mackrell, K., & Stevenson, I. (2009). Designing digital technologies and learning activities for different geometries. In C. Hoyles & J.-B. Lagrange (Eds.), *Mathematics education and technology: Rethinking the terrain (ICMI Study 17)* (pp. 47–60). New York, NY: Springer. http://dx.doi.org/10.1007/978-1-4419-0146-0_4.

NCTM. (2000). *Principles and standards for school mathematics*. Reston, VA: National Council of Teachers of Mathematics.

Ntenza, S. P. (2006). Investigating forms of children's writing in Grade 7 mathematics classrooms. *Educational Studies in Mathematics, 61*, 321–345. http://dx.doi.org/10.1007/s10649-006-5891-0.

PORDATA. (2013). *Students enrolled: by level of education and sex – Portugal*. Retrieved from http://www.pordata.pt/en/Portugal/Students+enrolled+by+level+of+education+and+sex-1005.

Shield, M., & Galbraith, P. (1998). The analysis of student expository writing in mathematics. *Educational Studies in Mathematics, 36*, 29–52. http://dx.doi.org/10.1023/A:1003109819256.

Statistics Portugal. (2013). Sociedade da informação e do conhecimento. Inquérito à utilização de tecnologias da informação e da comunicação pelas famílias 2013. *Destaque – Informação à Comunicação Social*. Retrieved November 6, 2013 from http://www.ine.pt/xportal/xmain?xpid=INE&xpgid=ine_destaques&DESTAQUESdest_boui=152142241&DESTAQUEStema=00&DESTAQUESmodo=2.

Chapter 3
Perspectives of Teachers on Youngsters Solving Mathematical Problems with Technology

Abstract This chapter offers the perspectives of teachers on youngsters solving mathematical problems with technology during the SUB12 and SUB14 mathematics competitions. Drawing on a series of interviews with teachers who have supported the participation of their students over several editions of the competitions, we identified what they see as the competitions' most significant features. The teachers spoke about the different kinds of support that are available to youngsters throughout the successive stages of the competitions, from the initial dissemination, to the online Qualifying phases, and lastly to the on-site Final. Based on their statements, the teachers say that they value the type of problems they characterise as challenging, real problems, appropriate for all students and useful as pedagogical resources. They make a distinction between such non-routine and extracurricular problems and the more school-like problems presented in mathematics textbooks. They are favourable to the use of technologies within the competitions, even when admitting initial difficulties that they nevertheless seemed to have overcome over the years. Some of these teachers enthusiastically describe how they sometimes integrated the competition problems into their class teaching and how they helped and encouraged students to use digital technologies for solving and expressing the solutions they submitted. The need to develop mathematical communication is seen as another challenge, and this, say the teachers, gave them the opportunity to explore different mathematical representations with their students. As a final point, several teachers highlighted the fact that youngsters' participation in the competitions was a motivating factor, contributing to their enjoyment of mathematics and feelings of inclusion in a community gathering many youngsters, parents and teachers around mathematical challenges.

Keywords Teachers' practices • Teachers' views on competitions • Participation • Mathematical problem-solving • Technology use • Challenge • Moderate challenge

3.1 Introduction

In recent years, there have been several activities in Portugal dedicated to mathematics with the aim of involving young people and increasing their appreciation for mathematics. With the mathematical competitions SUB12 and SUB14, these various activities have come to the attention of students to a great extent through their mathematics teachers. The growing impact of these types of activities in and out of school has sparked the interest of several researchers, and we can say that a new strand is emerging in research in mathematics education.

This chapter presents empirical data, mainly resulting from semi-structured interviews with several teachers, on the perspectives of mathematics teachers regarding their students' participation in beyond-school projects that are directly related to the use of technologies in mathematical problem-solving, namely, web-based competitions. More specifically, we present and discuss the perspectives of some teachers on youngsters' participation in the SUB12 and SUB14 mathematical competitions. Despite the fact that the competitions take place beyond school, teachers play a key role here. The teachers in their respective schools are the main links between the organisation of the competitions and the potential participants. As we show further below, the teachers of the young participants have had an influential role in the success of these mathematical competitions. Thus, there are strong reasons for a chapter dedicated to teachers in a project that has, as one of its research foci, the views held by individuals that directly or indirectly become engaged and strengthen the affective string connecting youngsters and mathematical problem-solving.

As we report in earlier chapters, the mathematics teachers of the 2nd and 3rd basic education cycles (grades 5–9) of the local schools are an essential bond between the University of Algarve and the elementary schools involved in these competitions. Through their teachers, students have had access to the publicising flyers that are sent annually to schools. In this sense, knowing the perspectives of teachers about this enterprise and the relevance that it can bring to their students' learning is of the utmost importance for research.

The teacher's role in the education of their students and, in particular, in their mathematics education is undeniable. In today's society, such a role is very demanding and requires different kinds of knowledge and the ability to perform a variety of tasks. Alongside our aim of understanding the perspectives of the teachers about these competitions, we wish to express our gratitude for the commitment, support and help from the teachers to their students. The efforts of all the teachers have gone far beyond the duties and skills required of them in the usual performance of their professional duties. This acknowledgment, that we want to emphasise in this book, is also shared by many parents and students in several statements that we have had the opportunity to collect over the years.

While we have been gaining awareness of the important role of teachers in these competitions, we have realised that different schools were creating different dynamics around this beyond-school project. Thus, our aim is to give a brief account of how these dynamics have developed in particular schools and with some teachers we deem to be laudable examples. This is of significance because it can make an important contribution to the future involvement of other groups of teachers in activities of this kind, which, as we know, are increasingly present in the current educational context and digital world.

First, we seek to identify the most outstanding characteristics pointed out by the teachers about these competitions, those which they see as an added value for their students and which lead them to stimulate and encourage the participation of the youngsters. Another aspect that has been discussed very often between the organisation, the teachers and the families is the question of helping the participants. As we have argued in previous chapters, our perspective is that offering help to the participants makes sense. We believe it is necessary to nurture problem-solving and that means of course providing help and support to young people in an initial phase until they can progress on their own. Therefore, it seems important to find out how teachers see the idea of help providing and how they put it into practice. Still on the subject of providing support, we believe in the importance of giving the participants a new opportunity to resubmit their solution. If we look at the classroom practices, this situation does not occur in a generalised way. Students who fail to solve a problem or a task correctly may develop a sense of failure and inability; usually, they get a negative assessment but are rarely given the opportunity to redo their work. Yet another aspect that distinguishes participation in the competitions from the classroom daily routines has to do with the issue of time. The time available to solve the problems in the competitions is more extended; participants have 2 weeks to produce their solutions to a problem, while in the classroom the time to solve a task is much shorter. This has clear reflections in the results, and it is quite noticeable that students can cleverly propose interesting and creative solutions as can be seen in several productions presented throughout this book. We are interested in understanding how the teachers see and interpret these differences, and we also want to get a sense of what they think about the proposed problems and the ways in which they present them to their students.

In addition, since the competitions promote the use of technologies for solving and expressing the solutions and require electronic communication, it is useful to know how teachers see and encourage the use of technology in problem-solving.

3.2 The Role of the Teachers in the Mathematical Competitions

The many mathematics teachers of almost 150 schools in the southern region of Portugal have had a decisive role in the success of this project. Every year they are requested to collaborate in disseminating these competitions, in their classrooms and among their students. This is the first task in which they are involved, but certainly one of the most crucial tasks. We collected various statements from participants that illustrate how the words of the teachers in the classroom are vital to motivate young students. Jenny, Rui and Mara are three youngsters living in different cities who participated from their 5th grade to their 8th grade in the two competitions. Their testimonies reveal the importance of the words of their teachers in motivating them to engage in these competitions.

> Jenny: The school teacher advised me. She said that an activity called SUB12 that was on mathematics was going to start and that anyone who wanted to participate could join. I wanted to participate and I took part in it.

Fig. 3.1 Image of the flyer of the mathematics competitions SUB12 and SUB14

Also Rui, a participant from another distant school, explained how he learned about this initiative. His mother stated that she will not forget the enthusiasm and brightness in his son's eyes the day he received the flyer with the information about the competition and the address of the website (Fig 3.1).

> Rui: I saw a poster in the school but did not care much. My math teacher explained, in a class, what it was and asked us if we wanted to participate. I came home with a great desire to participate.

A few kilometres away, Mara, who is a student from yet another school, recalled how she learned about the mathematics competitions SUB12 and SUB14.

> Mara: Oh… it was the teacher who said in the class that it was something that happens from January to June; there are problems we have to solve within the period of 2 weeks, and she said it was through the computer and gave us the essential information and passed on that… flyer… and that was it. I found it interesting to try… The teacher also said that in the end there would be a Final.

As shown by the words of the participants, the information given by their mathematics teachers in the classroom was very important. Although the advertising of this initiative takes place in several ways, including the posting of a poster in the students' room in each school, this appeal seems to be less effective than the discourse of the mathematics teacher.

However, the role of many teachers far exceeded the transmitting of information about the competitions. Many teachers took as their own the task to actively engage their students in this activity. Some chose to initiate students in solving problems posed during the first weeks, pushing them to participate later autonomously, while others followed the entire course of the competitions until the last stage, even being together with their students at the University of Algarve in the Final of each competition. One of our purposes in this chapter is to show how this path was followed and why the teachers supported their young students.

3.2.1 The Support of the Teachers: From the First Round to the Final

The way the teachers supported their students varied considerably from case to case. Some teachers just ensured the delivery of the flyers to their mathematics classes, as requested, leaving the participation in the competitions in the hands of the students; others took a more active role by engaging more intensively in the process of discussing the problems and helping and coaching students. The dynamics of each school is sometimes related to the involvement of teachers; therefore, we find schools in which adherence was quite broad, reflecting a certain attitude or policy of the school; in other situations, the involvement was more limited depending on only one or two of the more enthusiastic teachers.

Here, we offer a few examples that illustrate high levels of involvement of the teachers. In the course of each edition, we have met many teachers who might represent a variety of situations in the support and incitement given to their students for participating in the competitions. By personal contacts, e-mail exchanges or phone conversations, we were able to identify and select a few teachers that we decided to interview with the purpose of understanding the ways and reasons of their involvement. Therefore, we carried out several semi-structured interviews with teachers from different schools. These were also complemented with data collected over other episodes involving e-mails or phone calls from teachers themselves to the organisers of the competitions.

Let us first present Mr. Z, a teacher of 2nd cycle of elementary education (5th to 6th grade) who introduces these competitions to his students in 5th grade. Mr. Z is a dedicated teacher who seeks to encourage his students to participate in various initiatives, particularly related to mathematics. The SUB12 is one of the activities that he chooses to offer his students, encouraging them to participate, and furthermore whenever he can, he supports the organisation by collaborating in the Finals. Mr. Z explains his strategy in the first weeks of the competitions:

> Mr. Z: In the beginning, I bring the problem [to the class], explain what's going on, show them the webpage and explain what the competition is. I try to point out some of its advantages in order to encourage them and then explain how it works and try to support them too. From then on, although they do not use it much, I tell them that before sending the solution to the SUB12, they may send it to me and then if I think it's okay I recommend them to send it; if I think it's not good, then I give a suggestion; I don't give them the solution but I give a suggestion and tell them to check their work better. They don't use it a lot, but some do. That's in short what I do… And also in class, I read their work on the problems and if it's fine I say: "Look, you can send it". If it's not correct I may give a suggestion: "Look, see this part here", "Look, explain this better", so that they are the solvers; the idea is that they are the ones who solve it.

He added that this monitoring, in a more intense way, takes place in the early rounds, but then each student is responsible for their participation. However, he claims to be always available to help when needed, during class or outside the classroom. He suggests that students send him their solution by e-mail so that he could correct it or give suggestions about how to improve it.

Mr. Z: Typically, the first two or three problems, I solve them in the class… when I say solve them I mean that after the students try to unravel a problem, they sometimes take it home in order to think more about it… Later on, I end up solving the problems in class mostly to encourage the students. After they start progressing, I try to encourage them to solve a problem more on their own, with less help. Sometimes I select some [problems] that seem most appropriate for classroom work; there are others that I just leave to them, to those who participate in the competition…

Mr. H is perhaps one of the most enthusiastic teachers about the competitions. Since the first edition, he has closely followed the competitions, always very attentive. He has unfailingly kept track of each new problem as soon as it was published on the competition website. Mr. H prints it and then makes copies to bring to the classroom so that all students have access to the statement of the problem. Thus he seeks to give equal opportunity to all, ensuring that no one will be prevented from participating due to the lack of Internet access at home. If the organisation is delayed in releasing the new problem, Mr. H is the first to alert the organising team with care and attention. Sometimes he chose to contact the organisation to discuss the problem with great passion, the same passion with which he discussed it with his young students. In the early rounds of each edition, Mr. H ensures that all his students participate. If some students had difficulty in accessing the Internet or in sending the solution to the problem by e-mail, Mr. H always found a way. He considered all contingencies so that no student failed to participate. This teacher was always present at the Finals, being beside, and supporting, his young students. One child's mother made the following remark about the participation of her daughter in the Final:

Mother: All this is to the teacher's credit. It's him that inspires and mobilises them.

In another town, there has been another teacher who has also been a great supporter of the mathematical competitions SUB12 and SUB14 since the first edition. Mrs. A has spoken of these competitions and the participation of her students with great joy. As with Mr. H, Mrs. A has also been a regular presence in the Finals, sometimes collaborating with the organisation. Since the first edition, her school has in its development agenda the goal to involve and encourage their students to participate in the mathematical competitions SUB12 and SUB14. The importance that this school has devoted to the involvement of the students was reflected in the award of a diploma for their participation (Fig. 3.2).

Mrs. A: In our group [of mathematics teachers], we try to ensure that kids participate… participate…

She explained how the mathematics teachers committed themselves, from the very beginning, so that all students had the opportunity to participate.

Mrs. A: At that time, we still didn't have the rooms equipped with computers and with a video projector, so what we did was make copies of the problem, and I remember that we put them always in the library and in the teachers' room. We put copies in the students' room too, and we always made copies for all the teachers.

Mrs. A and her colleagues solved the problems before going to the class, and she confessed that often when they met in the teachers' room, they asked each other:

3.2 The Role of the Teachers in the Mathematical Competitions

Fig. 3.2 Image of a merit diploma that is awarded for participation in the competition

Mrs. A: Have you already solved the SUB14? Listen, what's the answer...?

She explained how she encouraged the participation of her students every year.

Mrs. A: I can tell you how we do it; in principle... this is how. The first thing, sometimes in a supervised study class and other times in a mathematics class, the problem is displayed and they all copy it, all of them... Often, I even deliberately pretend that I did not read the problem before... I enjoy going along with them and grasping... And I say: let's all solve it together. So some of them finish reading it and immediately say: Oh, Teacher, I know, I know, I know... And then I say: Yeah, okay, you're going to think again because I won't tell you just now if the answer is right... Then they go home and the next week we will see who thought about the problem and came to any conclusions. They often have difficulties... and then we discuss a little, each one tells their thoughts about the problem and how to explain it. I help a little and ask "So how did you think?" "Come here and tell the class..." and so they do. In the first problem, we help more; I usually go to those who have more difficulties: "So, where are you having trouble?"

At the same time as helping her students, the teacher encouraged them to do their own work. For example, after a discussion in a class, she told them to solve it at home, and the next week, before the submission deadline, she asked them about their difficulties and tried to help them to finish the problem. She was aware that many have parental support at home, but still she questioned them, as she explained:

Mrs. A: Then in the second part, I have no control, right? So they go home and have to find the answer, either they send it or not and either they get it or not. I mean there were several proposals and various justifications presented in class, right? Each of the students has to realise what the solution is; I don't give it, right? And then, when they get their answers, we review them and I might say, "So how is this solved? Does this have anything to do with some of the ways we considered? Can you remember?"

In all rounds, Mrs. A follows the same procedure: in the first week, they begin the discussion on the problem, later they take it home and the next week conclude the resolution and pose questions. The submission of the answers via e-mail is the responsibility of students at home.

3.2.2 *The Social Part of the Competitions: The Meeting at the Final*

In the introduction to *Challenging Mathematics In and Beyond the Classroom*, Barbeau (2009, p. 4) emphasises the importance of mathematical challenges in modern education:

> In many mathematical situations, children with their curiosity and mental agility are in a position of equality with adults. In particular, mathematical challenges become not only a way in which they can feel intellectually alive and productive, but also something that can be shared outside of their own age group.

To participate in the Final is something that motivates both the young participants and their teachers that often join them even if this event occurs at the weekend. Mrs. I reported that she escorted her students to the Final on several occasions. She noted that it was very important to be present in that event, in particular for the recognition of the work done by the participants:

> Mrs. I: Because it is much more interesting if they manage to get there [to the Final], I think so... for the child... to be there and then receiving the award and so on, this is very attractive to them... I think they like to go to the Final; they like to solve the problems correctly in the Qualifying and feel very happy to be there in the end.

The enjoyment and happiness are, as mentioned by several authors, some of the advantages identified in the research, resulting from the participation of young people in mathematical competitions (Amado, Ferreira, & Carreira, 2014; Freiman & Applebaum, 2011). Those affective aspects identified by this teacher are currently shown by research studies as important to the motivation of young people to the study of mathematics. On the meaning of the awards, Mrs. I added that these are also important for young people:

> Mrs. I: I think it's important that they receive awards because, after all, people like to be rewarded when they perform a task properly... but one thing is the prize and another is the value of the prize. Being awarded the prize... is the recognition. The value of the prize, I don't think it's important. So if it's one hundred euros or five hundred euros, I think it's not relevant. But of course they like to have a small prize, to be the 1st winner. Everyone likes it, but it's not really the importance of the monetary value itself. I don't think so.

The meeting that takes place during the Final was highlighted as a particularly important moment and one that makes youngsters aim for the Final. In this respect, another teacher stated that the participation in the snack break is a very important social fact for her students:

Mrs. D: The snack break. This is very important. The snack break is one of the things I realised that encourages students to participate in the Final. And believe me, a joke or not, some students said that one of the reasons to go to the Final was the snack break.

This social and cultural character of beyond-school projects is also referred to in the literature, and such dimensions of engaging with mathematics are increasingly necessary to promote a less harsh and cold view of mathematics among students and the community in general:

> Competitions and mathematics enrichment activities can be viewed as events that provide impetus for subsequent discussions among students (as well as among their teachers, friends and parents). From the viewpoint of acquiring new mathematical knowledge (facts and techniques) these "after competition discussions" might be as important as the preparation for and the competition itself. Many mathematicians owe a significant part of their knowledge to just such "corridor mathematics." From this point of view, the social programs organized after competitions provide additional importance. (Kenderov et al., 2009, p. 64)

Learning should be seen in a social perspective of participation in a community. The participation of young people in mathematical competitions with an inclusive character is an opportunity for youngsters to participate in a social activity, sharing the same context with their families, friends and teachers, a situation that is not usual in school mathematics. Some teachers seem to pay special attention to this as an important factor to encourage many students to participate. Indeed mathematics is sometimes seen as a difficult subject only accessible to a limited number of students and that rarely offers collective moments of socialisation. However, the Finals of these competitions allow a large number of students to meet and share with each other the pleasure of participating in such events. This seems to be an important issue for this age group; the enthusiasm and joyfulness that they demonstrate reveal that participation in activities related to mathematics can foster the enjoyment of mathematics.

The SUB12 and SUB14 Finals were alive and crowded with a large number of students, because the selection is not tight: participants just have to get eight of the ten proposed problems correct to reach the Final. Another very relevant fact was that the call for teachers, parents and families to attend was widely supported every year. This was an expression of appreciation for the role of this event by the community and young participants. In Figs. 3.3 and 3.4, we can see a teacher who was present in all the Finals, together with a group of his students (Fig. 3.3) and convivial moments during the snack break with youngsters and parents sharing the same lively atmosphere (Fig. 3.4).

3.3 Perspectives of Teachers About the Mathematical Competitions SUB 12 and SUB14

Research in the area of mathematical challenges for young students has proposed two distinguishable types of competitions in view of their more inclusive or exclusive characteristics (Protasov et al., 2009), although they should not be seen as

Fig. 3.3 A teacher and his students during a snack break at the Final of SUB12

Fig. 3.4 Convivial moments with youngsters and parents in the cafeteria during the snack break

incompatible. While the International Mathematical Olympiad (IMO) and other national competitions are exclusive competitions, we can find today examples like the SUB12 and SUB14 that are involving a wide range of students. Several academic studies (Jacinto & Carreira, 2011; Kenderov et al., 2009; Stahl, 2009a; Wedege & Skott, 2007) have revealed that such activities performed beyond school have important results in the development of problem-solving ability, mathematical communication skills, the emotional attachment of young people and families to mathematics and in discovering interesting uses of digital technologies to tackle mathematical problems. The purpose of the Problem@Web project is to contribute to this advancing knowledge thus reporting, in this chapter, on the perspectives of teachers who supported students in the school and in the classroom. We seek to know the advantages that teachers identify in the involvement of students in the SUB12 and SUB14 competitions, in particular with regard to the learning of mathematics.

We begin with the views of Mrs. I, a teacher working in the second cycle of education (grades 5–6), who since the first edition supports and closely monitors her young students as they take part in the competitions. She spoke of what she found most appealing when she began to get involved in this initiative.

> Mrs. I: I found it very interesting because it was the first time that students would be able to solve problems with enough time to think about it, to try to define several strategies if need be, etc… They would have enough time to look into the problem and simultaneously it was a period where the use of the computers wasn't yet very commonplace, isn't that so?! And so, for them, it was also the beginning of dealing with e-mails or even working with their computer for submitting their own solutions to the problems, so they were not used to doing things online, they used to do it all by hand.

This teacher emphasised various aspects that she recognised as important for the learning of her students: the issue of *time* for solving the problems and the idea of *development of technological skills*, in particular concerning the use of the electronic mail and of the computer to produce the answer to a problem.

Regarding the time factor, it should be stressed that students have 15 days to solve each problem and send the solution; this feature is a noticeable distinction between what problem-solving means in the classroom and in this beyond-school context. Usually, classroom time is limited to 50 or 90 min for solving tasks proposed by the teacher. The school puts students under a controlled time to solve a problem or whatever mathematical task. In contrast, in the competitions youngsters have their own time to think and develop a solution to a problem. This radically changes problem-solving to the extent that this activity is not compatible with a short time; rather, a real problem-solving activity requires that students have enough time to create a strategy, reflect on their work and express adequate reasoning. In fact, during each round of the SUB12 and SUB14 mathematical competitions, participants have several days to think and prepare their solution and may even reformulate and resubmit it. This option of extended time is reflected in the sophistication of some of the students' productions, as can be seen over the several chapters of this book. The teachers who followed their students throughout the competitions identified this significant difference, also stating that in the classroom this extended time is not possible under the pace at which they are obliged to comply with the official curriculum and because of the existence of examinations at the end of the 6th grade. As such, Mrs. I, for example, considered this opportunity for reformulation and resubmission that the competitions offered to be very important not only for her students but for all students, as she explained:

> Mrs. I: I really try to give time to all students and give them an opportunity… I help students to solve some problems; it is a way to integrate them, you know? It can't be all at once… If they work, if we give them a hand, they won't get stuck, they will advance; this is a good thing as all students may be integrated and will progress.

For this teacher, it was useful that all students had the opportunity to tackle the problems in the classroom, leaving the subsequent decision to participate or not in the competition to the individual student. This view aligns with Barbeau (2009) who argues that there should be freedom for students to decide on their participation in extracurricular activities; the success of participation of each student depends on the extent to which they really wish for it:

> … the mathematician or the teacher who introduces challenges into an environment must be aware of the particular circumstances. In the classroom, it is important to be as inclusive as possible, while the extracurricular activities were participation is voluntary, the educator cannot force anyone to take part and must select material carefully to ensure success. (Barbeau, 2009, p. 13)

The problems introduced in the SUB12 and SUB14 competitions were one of the most highly valued elements by several of the teachers interviewed. In this regard, Mrs. A said:

> Mrs. A: At school, we always embraced problem-solving. For many years, we had the problem of the fortnight. Then there were years when we were a bit tired, in which students did not react so much… So when your problems appeared, it relieved us from searching [for problems], because it's not easy to find problems that are meaningful… Because there are many problems that are only riddles and that are not rich enough to be explored in the mathematics classroom. I do not mean they can't have some interest… from the simplest to the most complicated, they always have some interest, but of course when compared with the SUB12 and SUB14. There the problems are very specific, have a very enriching approach to content which is always one of the aspects that we find important in problems…

For this teacher, who has always been a believer in problem-solving, the existence of the competitions made her job easier at the school in that it gave her a set of problems that she considers interesting and relevant to her students' work in the classroom.

As for Mrs. I, the quality of the problems is also the characteristic that most stands out in these competitions, saying:

> Mrs. I: To me, it's precisely the problems. In fact, I think there has been significant care in selecting very comprehensive problems, so they may have to do with the area of logic, others more with the area of geometry, others in which there are several connections and so on. So to me, it's all about the problems and the challenge that the problems pose to the students, right? So it's rather that part of the challenge itself… and as I say, most of them, as they are very diverse, allow developing many skills, so I think they are very rich. The competitive side is not everything, at least not to my students…

3.3 Perspectives of Teachers About the Mathematical Competitions SUB 12 and SUB14

This teacher also identifies several aspects that distinguish these competitions from others that are available to her students.

> Mrs. I: There are several differences… In fact they usually participate in the Mathematics Olympiad, they also participate in the Kangaroo, and they participate in the SUB12 and SUB14 competitions. In spite of all being, in the end, problem-solving contests, they are quite distinct. So, the Olympiad is really to try to determine which students are truly exceptional in problem-solving. Of course there are students who are most exceptional; others are not… Anyway you have to motivate students to learn the subject. The SUB12 differs essentially from others. As for me, this is the main question; it's very important to motivate students to mathematics. There is also another important aspect which is the involvement of parents themselves and siblings or relatives…

Mr. Z also referred to other competitions stating that it is not a question of labelling them as better or worse, but of understanding that they are diverse, have different purposes and are intended for different audiences.

> Mr. Z: I do not know many other competitions. This is a competition that I've already got used to; it doesn't mean that you think it's better or worse than others. For example, in the Olympiad, they get there, they have limited time to get the answers and it may be more difficult for some students to move forward… Okay, they have different characteristics, this doesn't mean they are necessarily better or worse, they just have different characteristics.

Mrs. I, and to some extent Mr. Z, differentiated the SUB12 and SUB14 competitions (in which their students regularly participated) from other competitions. They recognised, for example, that the nature of the problems posed in the SUB12 and SUB14 competitions was unlike that of other competitions, something that fits with what is known from the literature on curricular enrichment projects. In particular, and in contrast to exclusive and selective competitions such as the Mathematics Olympiad that aim primarily at detecting potential new talents for mathematics (Kenderov et al., 2009), inclusive competitions have other goals much more related to engaging a diversity of students in mathematical activities that suit their interests and capabilities. As argued by Stockton (2012, p. 51–52):

> The founding of team competitions and competitions for students from outside the special mathematics classes reflect a possible shift in the focus and purpose of competitions away from a strictly talent-search model to a more inclusive "enrichment" approach.

In fact, inclusive competitions like SUB12 and SUB14 are targeted at all students and are designed to motivate them for mathematics, to increase their problem-solving capability and to contribute to their learning of mathematics beyond their school learning.

The teachers we questioned believed that the problems posed in the SUB12 and SUB14 competitions were suitable for working with all students in the classroom. They regarded them as an important classroom resource, as mentioned by Mrs. A:

> Mrs. A: It is how the contents appear… almost without knowing that we are working on them… because often kids can retrieve different contents or use their everyday knowledge. That's what I think to be an important aspect of the problems; they are different from the problems that we all end up using at the completion of a chapter, that are basically an application and not much more, right? So, one of the really good things about the problems of the SUBs is that they are filling, say, a major gap. Students, for example, when they take the

final examination, they often can't get the answer to a question and ask themselves "Which content is this about?" And even ourselves, if we are given a problem, we may not know. Is this about similarity? Is this about the circumference? Is it trigonometry? Is it Pythagoras? We ourselves, we end up solving the problems that we come across in this way: "Oh, it's this content matter, and therefore we can solve it this way…" And besides, many teachers, regardless of how well they prepare for their classes, often don't show a great openness in terms of the capability to integrate something else when they are delivering a certain topic…

Mrs. A revealed awareness that solving problems in the classroom, often supported by the textbook, typically meant working on tasks aimed at reinforcing curriculum content. In fact, an issue greatly discussed in mathematics education research is to know how to integrate problem-solving in the teaching of various topics, sometimes as a starting point for learning new concepts, or as a context to work on representations and mathematical procedures, or as experiences promoting the development of mathematical reasoning and the activity of doing mathematics (Schoenfeld, 1991; Stanic & Kilpatrick, 1989). Still, mathematical problem-solving is a very important goal in the school mathematics curriculum in many countries. The relevance of mathematical problem-solving was influenced by several recommendations over time, recently reinvigorated by the frameworks of international large-scale studies such as PISA. In Portugal, within the mathematics curriculum for basic education published in 2007 (Ministério da Educação, 2007), problem-solving emerged for the first time as a competence to be developed in all the years of schooling. This guiding principle, following international trends, consists of establishing problem-solving as an organising axis of the mathematics teaching and learning process. However, this is not an easy task; one of the difficulties that many teachers face is that of finding real problems, as mentioned by Mrs. A or Mrs. I. The majority of the problems in textbooks are word problems, closed and routine, like most of the problems presented to students in the classroom. In this sense, the problems from SUB12 and SUB14 are somewhat different and new, as revealed by the data collected in interviews with the teachers.

One further aspect that many of the teachers we interviewed mentioned was the challenging character of the proposed problems. Throughout her interview, Mrs. A declared that not only did she feel challenged by the proposed problems but that all the mathematics teachers in her school also did. She confirmed that every 2 weeks, the conversation in the teachers' room included discussion about the problems.

> Mrs. A: Of course it's a challenge… and it still remains so for all of us. Some of us already have over 30 years of teaching…

This teacher displayed a great concern for all her students. She wanted to give to all students the same opportunity to participate in the mathematical competitions SUB12 and SUB14. This is a concern shared by the competition organisers—to involve a variety of students in the competitions and not only students with special liking for mathematics or the gifted ones. As described in Chap. 2, these competitions actually involved a diversity of students regarding their attainment in mathematics. While being concerned with all her students, Mrs. A identified the characteristics of the problems, which she describes as challenges, as a major benefit of these competitions.

3.3 Perspectives of Teachers About the Mathematical Competitions SUB 12 and SUB14

Another teacher, Mrs. P, also defined the problems of the competitions with a single word—challenge—and readily explained the reason for this attribute:

> Mrs. P: Because I think that every problem is a new challenge that they have to solve, they have to face. I think it's very interesting. Sometimes they look as really simple things... but sometimes it's hard, even for us. We look at the problem and think, "My God, how do I do this?" Then we start to think and... "Ah, this is very easy after all" and sometimes they are very simple things but we... Because they are different from those we usually do.

Again, we can notice the contrast with what is done in the classroom, the so-called content application problems in which the student knows beforehand the concepts to be used. It is the kind of problems that have to do with what the teacher did in recent lessons in terms of the delivered content. In this regard, we refer to an episode that happened with Mr. H, one of the interviewed teachers. Once one of the problems of the SUB12 was published, Mr. H called the organisation, raising a question:

> Mr. H: Look, the SUB12 problem which came out now, it requires knowing the greatest common divisor, but the kids haven't learned it yet.
> SUB12: No, it doesn't. This is a true problem! They can solve it; they don't need to know the greatest common divisor. They need to think and they will get there.

The idea that the problems of the competitions are different from the problems of the classroom is unanimous among all the teachers we interviewed. Mrs. P, who has several of her students participating in the competitions, acknowledged that these are not the typical classroom problems and as such students have to find different ways to solve them.

> Mrs. P: Because they have to strive to solve problems that are different, so they have to get different ways to solve them, not those usual ways they use in the classroom.

An interesting question that could lead to new studies has to do with something we find teachers claiming when seeing a solution of their students: *I did not expect this solution from this student*. We found that sometimes the young participants worked out solutions that surprised everyone and in particular their teachers. The classroom organisation and the limited time available may not allow students to show what they are capable of. Mrs. P recognises that it is not always the best students in the classroom who are the higher performers in the competitions, which sometimes causes astonishment in the teachers. In her opinion, this fact is related to a number of aspects that shape the school activity. From our point of view, it is fairly explained by the fact that the activity in the classroom includes a low diversity of activities. This contrast between the *in* and the *beyond* the classroom can reinforce the idea shared by various researchers on the need to diversify the type of tasks in the mathematics classroom (Ponte, 2007; Stein, Engle, Smith, & Hughes, 2008).

> Teachers who attempt to use inquiry-based, student-centered instructional tasks face challenges that go beyond identifying well-designed tasks and setting them up appropriately in the classroom. Because solution paths are usually not specified for these kinds of tasks, students tend to approach them in unique and sometimes unanticipated ways. (Stein et al., 2008, p. 314)

In other words, it is not enough to insist on drill and practice exercises to recap the content taught; instead, a rich activity with problem-solving is absolutely vital as it has been advocated for more than three decades.

> Mrs. P: The school gives greater value to the more rigorous and disciplined students. Maybe this is not very wise, but the truth is that it is so. Because there are deep-rooted patterns in our society and people are still governed by these very principles.

Creativity is not always easy to reconcile with a short time to solve a problem and with the characteristics of many problems that are presented in the classroom, usually associated with a particular mathematical topic, or with drilling and practising on calculations, as Mr. E emphasised:

> Mr. E: Most of the problems of the SUBs are formulated so that they are accessible to all students. I think that problem-solving is stimulating… Students often get to 5th grade and practically they have not solved any problem by then. They were not encouraged to solve problems; there isn't such a habit! Students get to the 5th grade and just want to do calculations. They just want to know: is this to add or to subtract? They were not encouraged to solve problems, to think…

To Mr. E, the problems posed in these competitions were within the reach of most students. This teacher believed that, at least in Portugal, there was no regular practice of solving problems in classes and in primary school students are too focused on performing calculations. Like the previous teachers, Mr. E associated the word challenge with the SUB12 and SUB14 mathematical competitions when asserting:

> Mr. E: I think it's a challenge. Challenge may be the right word, the more appropriate one. Because students are then challenged to solve problems, to show not only their knowledge but their reasoning, to compete with themselves, to have goals… to have the goal of solving that problem. To me it is basically challenge.

Mr. Z stated his point of view on the problems of these competitions, explaining how he sometimes used them in his classroom:

> Mr. Z: Sometimes there are problems, some of them from previous years, that you can use in class. There are always two points of view; it can be a problem that uses some knowledge that we are dealing with at the time, some content… but it can also be the other way, they must see that they do not need to use any stuff. I approve the two perspectives: it can be for them to practise the skills they have acquired, but it can also be a different thing and call their attention to the fact that it's not always like this, that you must think differently. I guess the two perspectives are valid.

And he referred to what he considered to be the functions of the problems in his teaching activity:

> Mr. Z: There is one problem from SUB12 that is one of the oldest, but I think it requires a good level of reasoning from them; they manage to discuss it and exchange ideas. It's not something that they can do right away, and this leads them to think a little more. I believe that many of the problems of SUB12 have this characteristic. I generally like them and think they are reasonable; if I didn't think they were appropriate, I would not use them with my students. I suppose the key is the thinking. There is another important point—we can use them both to introduce contents and to practise contents, and both are valid for me. But the most important thing is actually getting away from routine and to think, to think. Between

the routines there must always arise something different for them to wake up and see: "Hold on, this is different". I find this important.

Mr. Z, like other teachers, noticed that the mathematical activity in the classroom was fairly driven and needed to be more diversified for students to develop their skills further. In his classes, he sought to introduce problem-solving from a perspective that was in tune with the type of non-routine problems of the competitions. Moreover, he commented on his problem-solving practice with students in the class:

> Mr. Z: Generally, I propose the problems and give them time to think and solve. But before that I did some work on problem-solving with them. Thereon they should be aware—at least I would hope so—that when they read a problem and can't immediately find the answer, they will have to read and reread it several times, and they will have to try—all those stages, so to speak, of problem-solving. As I say: It's one more problem, only this time you have this competition; that's an extra motivation. So I usually let them move on, I give them some time for reading individually and for a few minutes I do not answer to anyone so that they take time and insist on reading, understanding and trying to do something. Sometimes I may give them a small hint, then there is a stage where they can exchange ideas between them and, eventually, when some have managed to solve it, there are students who propose solutions to the class and I usually ask the opinions of others and steer the discussion. Sometimes it may be carried on as homework, and in that case I only have the discussion the next day so that they realise that it's not a quick solution and that it requires more than a short time. And finally I end up validating the answers but I try… not to give them answers and I question even those that are well done. I want them to justify and to argue… "Let's see if this is really correct" to see if they are also confident of what they did.

The teacher believed that not all problems of the competitions should be linked to some specific curricular content and considered it to be a sensible option also because the SUB12 is devoted to students from both 5th and 6th grades.

> Mr. Z: (…) Because they [the problems] do not always appeal very directly to certain content knowledge, you notice that they are even designed for students of two consecutive grades. So, if they appealed only to school knowledge, it would seem that the 6th graders would be a little in advantage, maybe… There are parents who have already questioned me: "Oh my son is a fifth grader and will compete with the sixth graders!" I just tell them to take it easy!

The teacher also believed that the problems of the competitions were quite reasonable to the extent that they provided students with a different activity from what was typical in the classroom and he found benefits for the mathematics learning of his students, as highlighted in his words:

> Mr. Z: It is the nature of the problems that leads students to a form of work that is different to usual. Students who work on such problems learn; it improves their learning.

By definition, a challenge entails an element of difficulty and creates the need to overcome an obstacle. Barbeau (2009) elaborated on the idea of challenges in mathematics education, noting that the mathematical challenges deliberately encourage the recipient to seek a solution. A good challenge is one for which the individual has the necessary mathematical apparatus but is forced to deal with it in an innovative way; it generates emotions that are close to those of mathematicians when tackling authentic problems. Such mathematical challenges are usually seen by students as different from their school tasks, and even when perceived as intricate, they boost feelings of pleasure (Jones & Simons, 1999, 2000). Therefore, we propose a subtle difference between the idea of mathematical problem and the concept of challenging mathematical problem. A mathematical problem, usually conceived as a situation from which the initial and the final states are known but the process to move from the first to the last is not immediately available, has its grounds in the cognitive components of the problem-solving activity. In turn, a challenging mathematical problem includes a strong affective appeal by involving curiosity, imagination, inventiveness and creativity, therefore resulting in an interesting and enjoyable problem not necessarily easy to deal with or to solve (Freiman, Kadijevich, Kuntz, Pozdnyakov, & Stedøy, 2009).

The research has highlighted the need for attention to the degree of challenge of the problems posed to students, and the idea of moderate challenge has been proposed (Turner & Meyer, 2004). The propensity to perform a task seems to decrease in two situations: when expectations about getting success are very high (too easy) or very low (too difficult). The choice rests upon the situations in which the expected success is around 77 % (Turner & Meyer, 2004). Interestingly, it is reported that students who are given moderate challenges tend to reveal lower avoidance of help seeking. Moreover, the use of moderate challenge is most effective when help seeking is seen as legitimate and when the presentation and explanation of thinking are requested. These two aspects are clearly present in the SUB12 and SUB14 competitions and are essential categories of practices that, according to Turner and Meyer (2004), are intended to be challenging. This helps to understand why the mathematical competitions SUB12 and SUB14 are regarded as challenging environments where young people can expose and develop their skills in the field of mathematics.

3.4 Mathematical Communication: An Additional Challenge

The need for students to express their ideas and their reasoning in writing is an essential feature of the SUB12 and SUB14 mathematical competitions. Their solutions are only validated and accepted when conveying a clear account of the reasoning process that led to the answer. We recall that all proposed problems are followed by the prompt: "Do not forget to explain your problem-solving process!"

This requirement has become very stimulating also for the teachers engaged in helping their students. In the same way as the young participants, their teachers felt equally challenged and motivated by this requisite of the mathematical competitions SUB12 and SUB14. One of the interviewed teachers recognised that mathematical communication is a strong feature of these competitions. This teacher highlighted the uniqueness of the written component as a factor that initially can represent a greater demand for young students.

> Mrs. I: (…) Explaining the reasoning… that part of the mathematical communication, especially in writing… When orally they are fairly able to explain it, but when it turns to writing, it's very complicated for so many students, especially those who are very fond of mathematics; they like to write very little. So they like mathematics because with mathematics they do everything very briefly, right?! When we are moving from natural language, for

example, to mathematics, it gets really shorter… For something that is said in a large phrase, we get there and write three or four little things and that's it… the symbols. So they are used to mathematics being a lot shorter and when I want them to do the opposite (laughs), I mean, that they use everyday language, it takes a lot of work to write, to construct the sentences and explain… The sentence construction that's the least when expressing exactly the reasoning that they made… Yes, they write sentences but can they tell how have they reasoned? It's very difficult for them to explain how they have reasoned. Sometimes I say "Well, but how did you get this number?" and then they try to explain it, and I understand it… I mean, I understand because I'm a mathematics teacher. If someone else was trying to understand, that person wouldn't understand anything. In fact, I sometimes try to explain what they write to their classmates, and they don't get anything… and I know that the student was saying it correctly; it means that the reasoning was all very well, but it was poorly expressed.

The issue of mathematical communication is also referred to by Mrs. A. This teacher also explains the difficulties that teachers faced:

Mrs. A: One of our concerns was to provide guidance to the kids on how they could make a diagram, on how they could… send attachments, so even we had trouble suggesting to students how they could justify, because they didn't master it and it's difficult, isn't it?! Explaining everything… We are talking of some kids that are in the 5th or 6th grade… We ourselves had to develop as well when we wanted to help students to explain, give an adequate explanation of the situation and then I often needed to write it down. The kids were not used to explain, were not used to pick up a sheet and explain everything from beginning to end. So this forced us to find easy ways to explain things so that the paths they followed were clear to the people who read it.

The development of mathematical communication is one of the soft skills that the mathematics curriculum for basic education has been advocating for quite some time. Teachers found in these competitions a stimulus to help students in the development of mathematical communication, but were also challenged to find new ways to help students to express their ideas and mathematical reasoning. This has been recognised by Mrs. A who revealed that frequently she discussed with other mathematics teachers about strategies for working with students on the question of mathematical communication in the classroom.

Mrs. L, another teacher who followed the competitions, recalls that at various times she led students to elaborate diagrams for expressing their ideas and their reasoning as an alternative to the specific symbolic language of mathematics, knowing that not always the formal mathematical language is the better way to address a problem. She explained how she tries to help in her classes:

Mrs. L: (…) Diagrams, I value the diagrams. I advise them to consider: "Look, you can also get there this way". There are groups of students who already have more structured ideas and get there doing calculations. But so that others will not give up, I always say: "But you can do it with drawings, don't give up…" Because… if the student is able to schematise, then he catches the essential… he is able to solve any problem… and I think it helps in other disciplines. If students have the ability to analyse a table, to analyse a chart… when they are in sciences and are confronted with a graph, or a table… they will also know how to read it, and in history they will know too, and so on. I always say to my students that I don't just want the answer, I want what they have made, the reasoning they have made… For the person may have a wrong answer only because at some point they made a mistake somewhere… so, what's more important is the reasoning, to know why I'm doing this, this and this. If the answer is correct, okay, it's because there were no errors in the meantime.

In general, there seems to be some consensus among the various interviewees about the differences in communicating mathematically in the classroom and in these competitions. Mrs. I believes that the issue of mathematical communication in the competitions is more difficult than in the classroom due to the characteristics of the problems themselves. In the classroom, problem-solving tends to emerge associated with a particular mathematical topic which means that students immediately link the resolution of that problem to something that was recently handled in class. That does not happen in the competitions, as she explained:

> Mrs. I: It's this thing: they are on that topic… they are given a problem and they soon presume and they are impeded, isn't it?! They just think it to be about something that they had recently learnt, while with these problems this doesn't work; at the outset, those [problems]… are not integrated in anything at all, it's a new kind of problem for them to solve. So they will have to find the way they see fit and make diagrams or whatever they choose… It's different, it's different…

Mrs. I recalled an episode that happened a short time ago regarding the different ways of approaching the same problem by students of different grades. She noted that not everyone knew how to use multiplication of fractions but that did not stop them finding ways of representation that were useful for supporting the resolution:

> Mrs. I: [With the current problem] what is happening with my students in 5th grade is that they are doing it with geometric figures, using the knowledge they have because they have learnt about rhombi and they also know percentages. I don't know if any of them used percentages, or the size of angles, and so on. For example, the 6th grader who has already showed me a solution has solved it using only the operations with fractions but not the others [in 5th grade]; the others are doing diagrams and getting there…

Mrs. I added that whenever a new problem is published, she will soon access the webpage to solve it before their students. She admits that teachers have a tendency to use more formal procedures than those of the students, and she is happy by being surprised with solutions from her students she never imagined.

> Mrs. I: When a new problem gets out, the first thing I like to do is to solve it right away. Therefore when a problem gets out, I will go there and solve it at once. Of course we tend to solve it in a certain way; sometimes I solve it in two different ways. I say, "What if I try that way…?" But most of the time I am surprised by the students, because students actually end up having very different ways from those I expected. Sometimes those ways eventually correspond or are basically consistent with what I expected but often there are very different resolutions, and I'm wonderfully surprised. Indeed I have learned a lot from them.

3.5 The Use of Technology: The Sharing of Experiences Between Teachers and Students

The use of technology has been widely recommended for several years in the teaching and learning of mathematics, but its integration into the mathematics classroom has not been yet adequately accomplished. Despite the many efforts and recommendations, there seems to be a strong resistance to the use of technology in the teaching and learning of mathematics, regardless of the rest of the world being more and more dependent on the use of various technological resources. As the OECD (2012) report mentions, technology is everywhere, except at school. Not surprisingly, many of the teachers we interviewed revealed having some difficulties with handling digital technologies.

One of the purposes of the SUB12 and SUB14 competitions, in addition to the development of mathematical skills, was the development of technological skills associated with the use of computers to tackle problems and communicate mathematically on the ways to get the solution. These competitions can be considered pioneers in the way they promote an enduring connection between the target audience and the organising committee responsible for the publication of the problems and the feedback provision to the participants. From the very beginning, the two competitions reach the participants through the Internet, and electronic mail is the communication medium between each participant and the organising team. Visiting the website to access the problem is essential, but this requires having a computer, tablet or smartphone with Internet access. In 2005, when the competitions began, schools did not have the technological resources they currently have, which somehow hindered the work of the teachers. Mrs. A explained how she and her colleagues overcame those former difficulties, using printed copies of the problems that were being posted on the website.

The use of photocopies was the way to bring the problem to all the students and teachers of the school, in the early years of the competitions. But it was not just the lack of resources in the schools; the obstacles also included experiencing difficulty with the use of the computer by students and by teachers. One of the requirements for participation was that each student had a valid e-mail address in order to send the answers and receive feedback from the organisation. Many of the students who started their participation, especially in 5th grade, did not have an e-mail address at the time. Some teachers took on the task of creating e-mail accounts for their students; others asked the parents to do it or asked for support from the administrative staff of the school. Mrs. A briefly described how this took place in the first editions of the competitions:

> Mrs. A: I remember that in the early years I didn't even handle the computer, I didn't even know how to create e-mail accounts. I didn't know... It had to be a staff member from the secretariat or a colleague who created the e-mail accounts for kids participate because we didn't know; we were all insecure... with over 30 years of profession...

The teacher admitted that at the time the competitions began, her skills on the use of technologies were insufficient to meet the required needs. Many of the teachers of her school had a long teaching experience but had few technological skills as their training never included this area of knowledge. However, we should point out that this fact was not an impediment to them in embracing the challenge and encouraging young students. On the contrary, they bypassed all the difficulties as seen in Mrs. A's words. For example, to create the e-mail accounts, they requested support from colleagues, which shows the commitment of the whole school to the involvement of the students in the competitions, something we should mention and praise.

Over the years, these difficulties have been settled, particularly with the rise of the National Technological Plan that in addition to equipping schools with technological resources also promoted the training of Portuguese teachers and the development of some practical skills on the use of technological tools. At the same time, hiring young teachers with more advanced technological competences also contributed to helping the more established teachers in computer use, as described by Mrs. A:

> Mrs. A: Now there are many young teachers being hired, but at that time we all had about 30 years of teaching, or close to that, and had little knowledge of technology, so that was all a novelty. And we also counted a little on the families who already had an e-mail account… So that was more or less how things worked…

It was not only in the school where Mrs. A taught that difficulties were felt. Mrs. I also acknowledged that it was not easy in her school and explains what her role was in the first years. This teacher, as she said in her interview, thought it was very important that students began to learn how to write their own e-mails and solve the problems on the computer to send them to the competition.

Mrs. I already showed a certain ease with using the computer, back in 2005. The organising team of the competitions began receiving several solutions from her students through the teacher's own e-mail account. This never caused any problems to the organisation, because the resolutions had been made by her students, sometimes with paper and pencil and after digitally scanned, with others sent as attached files. A few years later, when we interviewed this teacher, we had the opportunity to learn in detail about her work and her dedication in enabling the participation of her students.

> Mrs. I: I went with three [students] at a time [to the computer] for them to create their own e-mails and they all got e-mail accounts. But then there was another problem in the school, often the Internet did not work… At this stage, the Internet did not always work, and some kids had no Internet at home and therefore it ended up being me who sent the solutions that they produced… It was like that. Now most of the youngsters already have their own e-mail… sometimes what they don't have is the Internet… for financial reasons… In the early years of the SUB12, many youngsters did not have their own e-mail. Now if we ask about it in a class, we find only one or two youngsters who don't have their own email.

The emergence of technological resources in schools has brought some interesting changes that are relevant to consider. From the moment the classrooms began to be equipped with video projector and computer, Mrs. A and her students began to see, in the classroom, the scores' tables and the published solutions of some of the participants. As reported by the teacher, these moments with her students enabled the sharing of knowledge between teachers and students, as she recalled:

> Mrs. A: We open the website of the competitions, and we display the table and the solutions that are visible there. And then one of the things they really like is when they see… It's a joy to see their names in the answers and even in the table of scores, it's a joy because they like to show up, right, when we project it… In the class, I project everything on the screen. When the answers come [the teacher refers to the solutions selected by the organisation that are published on the website], we all look at them…

3.5 The Use of Technology: The Sharing of Experiences Between Teachers and Students

The teacher told us, quite amused, how she has learned from her young students to search the table and to use some computer tools. She added her memories of an episode where students actually suggested to her to use Excel to solve one of the problems:

> Mrs. A: In fact, once there was this problem and I wanted to explain something, and at some point one of the students asked me, "But can't we use Excel?" And I couldn't do it in Excel, but then there was one of the kids who came to the computer and did it!

Mrs. A recognised that everything is easier with regard to access to technologies both at school and at home, but she acknowledged that:

> Mrs. A: It's still nice, in the classroom, when they can see their names in the table and when there is [on the webpage] the resolution of one of them, the technology allows that, doesn't it?

Another teacher who taught in grades 7–9 said that in her school she had the chance to use a computer room to teach her mathematics lessons and periods of supervised study, allowing student's access to the Internet to see the competition webpage:

> Mrs. S: My students had the opportunity to go directly to the SUB12 and SUB14 webpage to read the description of the problem and see the resolutions of other problems that were selected by the coordinating team.

This teacher encouraged her 8th grade students to solve the problems in class and, if possible, to send the answers still during the class, as some of her supervised study lessons were dedicated to addressing the problems from SUB14. She explained that initially the students solved the problems with paper and pencil though they had a computer at their disposal.

> Mrs. S: Initially most of the students solved the problems with pencil and paper; however, as we had no scanner in the room and they had to send the solution by e-mail, they typed their resolutions directly in the reply window—if it was short—or in a word document. In this case, often the resolution sent wasn't quite the same as the one made with pencil and paper. Some students tried to use images where they put captions, others tried to use variables and carried out calculations and used mathematical language. In general, they tried to send a more formal resolution than the one they had originally done.

The teacher also thought that the use of the computer had become very important for student learning, enabling the development of many different skills:

> Mrs. S: Using the computer proved to be important for students in different aspects: in the mathematisation process, in learning and using digital representations in the problem-solving process and other skills on computers.

Mrs. S, like other teachers of young participants in the competitions, was a great believer in using Excel and as such sought to implement the use of this tool with her students in the classes. Some students were very keen about Excel and used it quite often.

> Mrs. S: Throughout the competitions ,I encouraged the students to use the spreadsheet in solving numerical or algebraic problems and thereafter some of the students began to use the computer as a tool to solve problems and not only as a means to express their answers

and send them. On the other hand, the use of computers has made the spreadsheet known to the students and allowed students, less used to it, to submit attachments.

Interestingly, since the first edition, we found several participants who showed great appreciation for the use of Excel; sometimes they resorted to Excel without any apparent reason, almost as if they were using a Word document, as illustrated in Fig. 3.5, which was basically produced using the insertion of shapes into an Excel sheet.

Another teacher claimed that the fact the competition takes place over the Internet only has advantages and considers that at the present time it would be meaningless if it were otherwise.

> Mrs. P: Well, because it's a practical way; it's a medium that kids usually use and master well and that they like. Maybe if it were on paper, we would have more difficulty, because we must always send the solutions, it would be more complicated and nonsensical.

We believe that technology can make an important contribution to the learning of mathematics. Initially, as we have seen, either teachers or students felt some difficulties, but the use of technology has become more readily accessible, particularly to students, who quickly become attached and learn to take advantage of its potential. Many of the solutions produced by the students over time reveal a level of effectiveness and creativity that would not be possible without the use of technology. Nevertheless, it is interesting to note how several teachers, with different levels of technological skills, opened their minds to taking advantage of the computer by supporting youngsters throughout the competitions. We can only praise the role of these teachers who have given us really significant testimonies as they are examples of how the teacher can offer students a mathematically stimulating and fruitful environment.

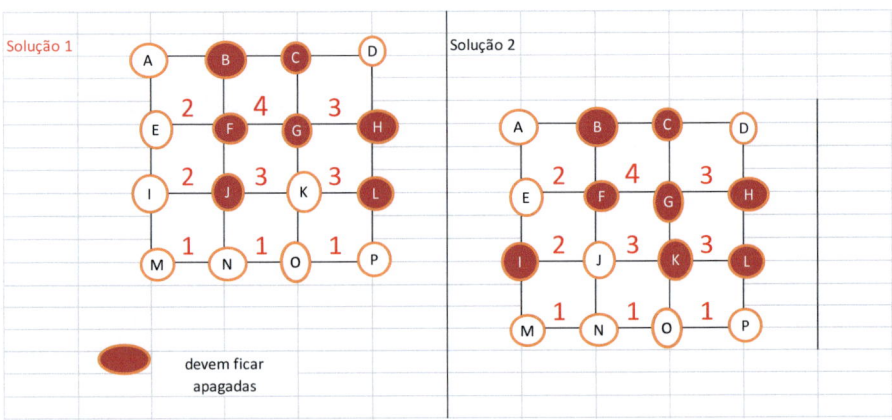

Fig. 3.5 A solution from a group of two girls to problem #3 of SUB12 (edition 2010–2011)

3.6 Overview and Conclusion

Throughout this chapter, we have reported on the involvement of teachers in the participation of their young students in the mathematical problem-solving competitions SUB12 and SUB14. Teachers have played a very important role in the dissemination of these competitions. Without their contribution, it would have been impossible to have achieved the massive participation of youngsters observed throughout the various competition editions. The evidence we collected show that this invitation to participate respects the freedom of every youngster to choose to participate or not in the competitions, which meets the recommendations offered by Barbeau (2009)—no youngster should be forced to participate in beyond-school activities. Nevertheless, it seems clear that teachers know how to motivate and inform students and even select, according to their principles and perspectives, the competitions that they consider most appropriate to youngsters. Further, this invitation appears to be made in a very effective manner taking into account the number of participants achieved. Although the mathematics teacher takes most often the role of promoter and motivator for participation in this activity, several teachers have an enduring engagement, following the course of this competition throughout the Qualifying and also, in some cases, being present, alongside the students, in the Final which takes place at the University of Algarve, in June each year. Several of the interviewed teachers used, in their classes, the problems proposed in the various rounds of SUB12 and SUB14. The resolution of the problems of the competitions in the classroom is indeed a strategy followed by many teachers. In fact, some indicated that the problems from the competitions are an interesting educational resource for their teaching practice, saving them the effort to seek and find good problems elsewhere. In some cases, introducing those problems in the classes was also the grounds to promote the use of technology, in different modalities, in mathematics teaching.

Various reasons were mentioned by the teachers to justify their commitment to this project: first, they mentioned the problems themselves as an educational resource that was valued as challenging and real problems that were different from the traditional problems posed in the classroom, namely, the problems taken from the end of each chapter in the textbook. The SUB12 and SUB14 problems were seen by the teachers and their students as distinct problems from the specific content-reinforcing problems aimed at practising recently taught topics. The teachers found that the problems were an important means of both introducing new mathematical topics and consolidating topics already studied, given that the problems were not intended to fit any specific content and therefore could lead students to make connections between different curricular topics and several mathematical ideas.

In addition to the challenging nature, other characteristics were also pointed out, in particular, the fact that mathematical communication is likely to be developed within the context of solving and expressing mathematical problems. Teachers recognised that, like their students, they were challenged to improve their skills and their communication strategies. The requirement set by the competition of sending the solution and the reasoning involved in solving the problem led teachers to devise and use different strategies in the classroom. Teachers felt the need to foster the development of several representational forms, namely, the use of drawings, diagrams or tables. Finally, the teachers all agreed that their students' participation in this type of inclusive competition promoted their fondness and enjoyment for mathematics as argued by several authors (Amado et al., 2014; Freiman & Applebaum, 2011). Signs of happiness and joy from youngsters participating in the SUB12 and SUB14 Finals were clearly revealed and acknowledged by the teachers who saw it as another valuable attribute of these competitions. This was also reflected on their willingness to collaborate, attend and join their students in the social and public moments, where parents and relatives are also included, such as the snack break or the awarding ceremony.

We conclude this view on the role of teachers, and their positioning in relation to proposals that extend beyond the classroom, by emphasising the importance they attached to such projects as sources of learning and accomplishment of their students. To a large extent, the statements of the teachers we heard echo the words of Stahl (2009b), referring to a vision of his Virtual Math Teams project:

> Students learn math best if they are actively involved in discussing math. Explaining their thinking to each other, making their ideas visible, expressing math concepts, teaching peers and contributing proposals are important ways for students to develop deep understanding and real expertise. There are few opportunities for such student-initiated activities in most teacher-led classrooms. (p. 24)

After having carried out, over two chapters, a rich description of the youngsters who engaged in SUB12 and SUB14 and the perspectives of their teachers who promoted and stimulated their activity as mathematical problem-solvers with technology, we move forward to a theoretical conceptualisation around the concept of solving and expressing and the inseparability between technology use and mathematical approaches to problem-solving. Our purpose is to develop theoretical tools to inform a sharpened analysis of how these youngsters made their ideas visible with and through digital technologies and particularly of how they expressed powerful conceptual models of specific problems of the competitions by ingeniously utilising digital tools of their choice.

References

Amado, N., Ferreira, R., & Carreira, S. (2014). *A relação afetiva dos jovens e suas famílias com a matemática e a resolução de problemas no contexto de competições matemáticas inclusivas.* Unpublished Monograph of the Problem@Web Project.

Barbeau, E. (2009). Introduction. In E. J. Barbeau & P. J. Taylor (Eds.), *Challenging mathematics in and beyond the classroom: The 16th ICMI Study* (pp. 1–9). New York, NY: Springer. http://dx.doi.org/10.1007/978-0-387-09603-2_1.

Freiman, V., & Applebaum, M. (2011). Online mathematical competition: Using virtual marathon to challenge promising students and to develop their persistence. *Canadian Journal of Science, Mathematics and Technology Education, 11*, 55–66. http://dx.doi.org/10.1080/14926156.2011.548901.

References

Freiman, V., Kadijevich, D., Kuntz, G., Pozdnyakov, S., & Stedøy, I. (2009). Technological environments beyond the classroom. In E. J. Barbeau & P. Taylor (Eds.), *Challenging mathematics in and beyond the classroom: The 16th ICMI Study* (pp. 97–131). New York: Springer. http://dx.doi.org/10.1007/978-0-387-09603-2_4.

Jacinto, H., & Carreira, J. (2011). Nativos digitais em atividade de resolução de problemas de matemática. In A. Andrade, C. Lajoso, J. Lagarto, & L. Botelho (Eds.), *COIED 2011 – 1.ª Conferência Online de Informática Educacional* (pp. 69–77). Lisboa: UCP. http://www2.coied.com/2011/ebook/.

Jones, K., & Simons, H. (1999). *Online mathematics enrichment: An evaluation of the NRICH project.* Southampton: University of Southampton. http://eprints.soton.ac.uk/11252/.

Jones, K., & Simons, H. (2000). The student experience of online mathematics enrichment. In T. Nakahara & M. Koyama (Eds.), *Proceedings of the 24th Conference of the International Group for the Psychology of Mathematics Education* (Vol. 3, pp. 103–110). Hiroshima, Japan. http://eprints.soton.ac.uk/41271/.

Kenderov, P., Rejali, A., Bussi, M. G., Pandelieva, V., Richter, K., Maschietto, M., et al. (2009). Challenges beyond the classroom: Sources and organizational issues. In E. J. Barbeau & P. Taylor (Eds.), *Challenging mathematics in and beyond the classroom: The 16th ICMI Study* (pp. 53–96). New York, NY: Springer. http://dx.doi.org/10.1007/978-0-387-09603-2_3.

Ministério da Educação. (2007). *Programa de Matemática do Ensino Básico.* Lisboa: Direção Geral de Educação.

OECD. (2012). *Connected minds: Technology and today's learners, educational research and innovation.* Paris: OECD Publishing. http://dx.doi.org/10.1787/9789264111011-en.

Ponte, J. P. (2007). Investigations and explorations in the mathematics classroom. *ZDM: International Journal on Mathematics Education, 39*(5–6), 419–430. http://dx.doi.org/10.1007/s11858-007-0054-z.

Protasov, V., Applebaum, M., Karp, A., Kasuba, R., Sossinsky, A., Barbeau, E., et al. (2009). Challenging problems: Mathematical contents and sources. In E. Barbeau & P. J. Taylor (Eds.), *Challenging mathematics in and beyond the classroom: The 16th ICMI Study* (pp. 11–51). New York: Springer. http://dx.doi.org/10.1007/978-0-387-09603-2_2.

Schoenfeld, A. H. (1991). What's all the fuss about problem solving? *ZDM: International Journal on Mathematics Education, 91*(1), 4–8.

Stahl, G. (Ed.). (2009a). *Studying virtual math teams.* New York, NY: Springer. http://dx.doi.org/10.1007/978-1-4419-0228-3.

Stahl, G. (2009b). The VMT vision. In G. Stahl (Ed.), *Studying virtual math teams* (pp. 17–29). New York, NY: Springer. http://dx.doi.org/10.1007/978-1-4419-0228-3_2.

Stanic, G., & Kilpatrick, J. (1989). Historical perspectives on problem solving in the mathematics curriculum. In R. I. Charles & E. A. Silver (Eds.), *The teaching and assessing of mathematical problem solving* (pp. 1–22). Reston, VA: NCTM and Lawrence Erlbaum.

Stein, M., Engle, R., Smith, M., & Hughes, E. (2008). Orchestrating productive mathematical discussions: Five practices for helping teachers move beyond show and tell. *Mathematical Thinking and Learning, 10*(4), 313–340. http://dx.doi.org/10.1080/10986060802229675.

Stockton, J. C. (2012). Mathematical competitions in Hungary: Promoting a tradition of excellence & creativity. *The Mathematics Enthusiast, 9*(1 & 2), 37–58. http://www.math.umt.edu/tmme/vol9no1and2/TME_2012_vol9nos1and2_pp1_232.pdf.

Turner, J., & Meyer, D. (2004). A classroom perspective on the principle of moderate challenge in mathematics. *Journal of Educational Research, 97*(6), 311–318. http://dx.doi.org/10.3200/JOER.97.6.311-318.

Wedege, T. & Skott, J. (2007). Potential for change of views in the mathematics classroom? In D. Pitta-Pantazi & G. Philippou (Eds.), *Proceedings of the Fifth Congress of the European Society for Research in Mathematics Education* (pp. 389–398). Larnaca, Cyprus. http://www.mathematik.uni-dortmund.de/~erme/CERME5b/.

Chapter 4
Theoretical Perspectives on Youngsters Solving Mathematical Problems with Technology

Abstract Given that solving mathematical problems entails developing ways of thinking and expressing thoughts about challenging situations where a mathematical approach is appropriate, this chapter unveils a theoretical framework that aims to guide a better interpretation of students' capability to solve mathematical problems with digital technologies, in the context of online mathematical competitions. The main purpose is to provide a way of understanding how students find effective and productive ways of thinking about the problem and how they achieve the solution and communicate it mathematically, based on the digital resources available. By discussing several theoretical tools and constructs, a theoretical stance is developed to conceptualise problem-solving as a synchronous process of mathematisation and of expressing mathematical thinking in which digital tools play a key role. This theorisation draws on the role of external representations and discusses how a digital-mathematical discourse is used to express the development of the conceptual models underlying the solution. In this conceptualisation, a symbiotic relation between the individual and the digital tools used in problem-solving and expressing is postulated and outlined: the inseparability of humans and media sustains the idea that students and tools are agents performing knowledge in co-action, while approaching mathematical problems. Looking at the solution to a problem is seeing a fusion of the solver's knowledge and the tool's built-in knowledge, rather than an aggregate of both or a complementarity between them.

Keywords Problem-solving and expressing • Conceptual model • Problem-driven conceptual development • Mathematical thinking • Mathematisation • Digital-mathematical discourse • External representations • Humans-with-media • Co-action

4.1 The Theoretical Stance

In this chapter, we provide the theoretical perspectives which frame the study of a specific and relatively new phenomenon, that of students solving mathematical problems with the digital technologies of their choice. With that purpose in mind, the theoretical tools and constructs that are reviewed and discussed lead to creating new constructs which are expected to guide a better interpretation of what young people are able to do in a digital communication context (in our case, finding the solution to a challenging word problem and its explanation within the scope of an online mathematical competition). Our aim is to understand how they find effective ways to achieve the solution of a problem and to communicate it mathematically, based on the digital resources they have at their disposal in their daily life, most cases in their home environment but also in school, including in the mathematics classroom.

The background of the study is centred on a particular theoretical stance from which we consider the *problem-solving process as a synchronous process of mathematisation and of expressing mathematical thinking*; such perspective is additionally supported by two specific research backings: (a) the role of external representations in problem-solving and expressing and (b) the symbiotic relation between the individual and the digital tools in a problem-solving and expressing technological context.

As illustrated in the previous chapters, the problems that are regularly proposed in the mathematical competitions SUB12 and SUB14 represent moderate mathematical challenges and share the attribute of allowing the construction of several conceptual models. Our Problem@Web project, therefore, focused on the extent to which young participants use conceptual models to approach a mathematical problem. This is especially important when problems are intentionally open to multiple alternative models.

Those who defend the idea of problem-driven conceptual development have argued that problem-solving needs to be treated as a way to put students in a process of developing conceptual models which include, within them, mathematical concepts and models of real situations and the thinking frameworks for a certain class of problems (English & Sriraman, 2010; Lesh & Doerr, 2003a). The problem-driven conceptual development theory dates back to the days in which Lesh and his collaborators stressed, for example, the relevance of the applied problem-solving as a fertile ground for conceptual models. A central aspect of this perspective is that the conceptual models organise themselves and develop, to a great extent, around lower-order concepts, many of which can be materialised through physical and manipulative objects and materials. As Lesh explains:

> These conceptual models contain ideas from which a maximum number of lower order concepts can be derived and which can be applied in a maximum number of situations. (Lesh, 1981, p. 245)

In the competitions that are the basis for our research, problem-solving is not confined to giving students the opportunity to apply what they have learnt at some point earlier. Their problem-solving is rather about them recognising and treating a problem by means of some mathematical approach that proves adequate to solve it and that allows various mathematical explorations and generating mathematical knowledge. This has to do with understanding problem-solving as a ground to develop mathematical understanding (or equivalently to promote the development of mathematical concepts) and even as a more transversal mathematical capability—that of knowing how to deal with mathematisable situations (Doorman & Gravemeijer, 2009;

Francisco & Maher, 2005; Gravemeijer, 2002, 2007; Gravemeijer, Lehrer, van Oers, & Verschaffel, 2002; Lesh, 2000; Lesh & Doerr, 2003a, 2003b; Lesh & Harel, 2003; Lesh & Zawojewski, 2007; Lester & Kehle, 2003; Santos-Trigo, 2004, 2007; Zawojewski & Lesh, 2003).

Based on this theoretical background, we intend to isolate a range of essential concepts which inform the direction of this study. We therefore establish a starting point: *mathematical problem-solving means ways of thinking and expressing thoughts about challenging situations where a mathematical approach is appropriate, even if the problem-solver may not recognise such thinking as being a typical mathematical activity or may not draw on school mathematics knowledge.* In a way, we assume that lower-order concepts are key anchors in young learners' problem-solving. Following this first assumption, we look for ways to theoretically reflect on what is involved when youngsters are using digital tools to think about and tackle mathematical problems. We recognise that digital tools provide affordances to support thinking about problems in terms of different models, including models that students would not immediately produce without these tools. The identification and understanding of these new models is one of the main motivations for the development of a theoretical framework sustaining the research on students' problem-solving supported by their use of technology. We are interested in the relationship that takes place between the solver and the digital tool as a potential form of addressing youngsters' digital skills as actual mathematical ways of thinking, modelling and expressing ideas and concepts. Abundant imagery based on the use of pictures created with drawing tools or the possibility of using images available on the Internet, the construction of tables in Word and Excel, the production of active calculations in Excel formulas, the precision drawings and many other representational possibilities of GeoGebra are among the set of elements that we want to theorise in discussing the idea of solving and expressing with digital tools. Another focus of the theoretical development refers to the mathematical thinking verbalised in the solver's narrative involved in the solution of a problem. The SUB12 and SUB14 are characterised by requiring forms of narrative presentations of solutions, which promote important forms of exploratory and explanatory discourse aligned with mathematical thinking. The narratives highlight (for students and researchers) the models used by the students in devising productive ways of thinking about the given problems. Therefore, the notion of mathematisation, as a basis for the search of productive models to solve problems, needs to be complemented by the expression of mathematical thinking with digital tools and thus leads to the idea of an emerging digital-mathematical discourse.

4.2 Problem-Solving as Mathematisation

Problem-solving considered as an activity which prompts the development of conceptual models is a way of describing the process of mathematisation or modelling (in the sense of building conceptual models). As noted by Gravemeijer (1994, 1997), this is actually the great inspiring breath of Realistic Mathematics Education (RME): to face mathematisation as an activity which requires, first of all, to organise and to structure a situation. As such, "Mathematizing, which stands for organizing from a mathematical perspective" (Doorman & Gravemeijer, 2009, p. 200) is the core activity that Freudenthal proposed to characterise the guided reinvention of mathematics while students engage in meaningful problems. The consequence of such activity is, in a first sense, a conceptual model that is a way to deal efficiently with the problem and, in a second sense, a mathematical model or a *mathematical skeleton* that is the basis of the solution to the problem.

However, in this type of activity, the model which arises from the mathematisation process does not always have the level of sophistication that we would immediately call a mathematical model. The conceptual model has, many times, an unsophisticated appearance, an informal tone and a situation-framed aspect. As Gravemeijer (1997) explains, in a well-chosen example, students' informal models are context-specific and often contain details and inscriptions that are tied to the concrete situation and reflect their ways of acting on it rather than being just superfluous and ornamental minutiae.

Echoing the example presented by Gravemeijer (1997, p. 394) about a student's solution to a problem about the division of sweets by three children—where "the informal character of the model is underlined by the fact that the student even tried to copy the children's portraits faithfully"—we offer a solution presented by a sixth grader participating in SUB12 to a combinatorial problem using the computer (Fig. 4.1). The problem mentions a girl, Isabel, who is departing to live abroad and says goodbye to her six best girl friends; at the airport she takes a photo of each pair of girls, including herself, and the question is to find the number of photos taken.

The solution to the problem was given in a PowerPoint slide, and it involved the use of colour, written words, calculations, diagrammatic arrangements and digital drawing. In particular, the faces of the seven girls mentioned in the problem were drawn and designed with Paint and then copied and pasted to small squares suggesting actual photos. Each girl's face is different and has several distinctive details, namely, the colour and the look of the hair. As Gravemeijer points out, this is a good example of how "at the referential level, the model refers to the situation sketched in the problem statement" (1997, p. 394). Nevertheless, as Lesh (2000) stresses, conceptual models only work in full when they are expressed through a variety of external representations, many times, simultaneously. These representations can be wide ranging, from written symbolic forms to graphical forms, from real and concrete models to the metaphors based on experiences, or yet to the dynamic and touchable images on computer screens.

When we observe the digital solution presented to the problem, we can recognise the conceptual model which gives sense to the situation and to the way of achieving the solution: list all the elements, set the first element and consider all the pairs that can be made with this fixed element and then remove this element, set the second element and count all the pairs that can be formed with the fixed element and so on until the last pair.

The conceptual model underneath shows the process of listing, counting and calculating, which can later be generalised and translated into a more formal and symbolic mode:

4.2 Problem-Solving as Mathematisation

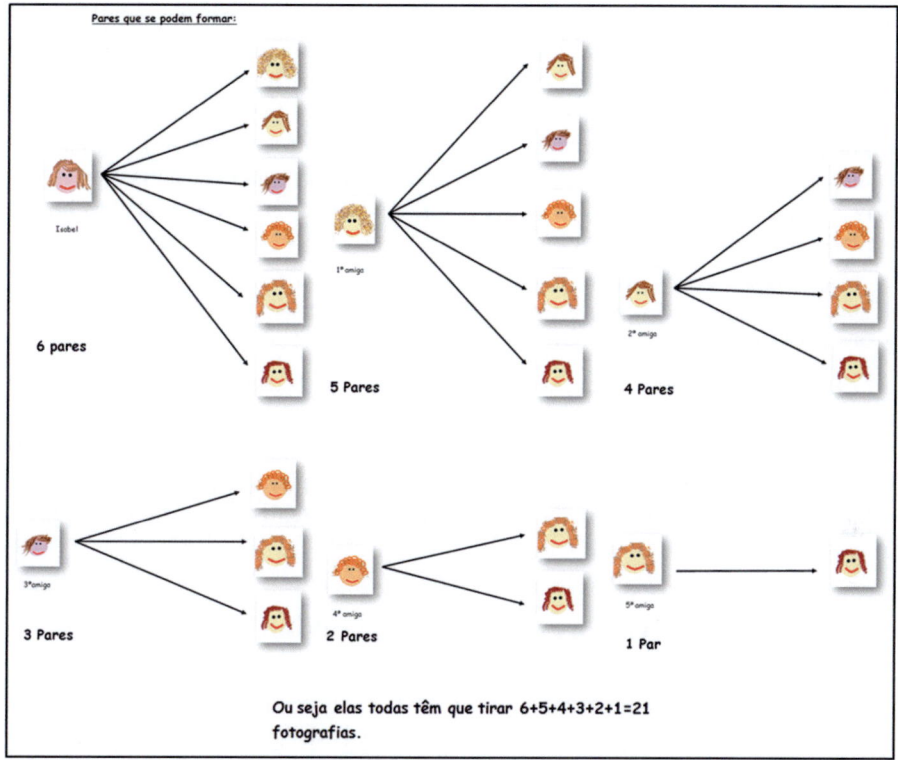

Fig. 4.1 Example of a solution produced with the computer by a sixth grader to a combinatorial problem proposed in the competition SUB12

Given n individuals, the number P of pairs of individuals that can be made for the complete set of photographs is given by $P=(n-1)+(n-2)+\ldots+2+1$.

An instantiation of this conceptual model appears in the account given on the Wolfram MathWorld webpage to explain the solution $C(n,2)=n(n-1)/2$ to the well-known handshake problem:

> To see this, enumerate the people present, and consider one person at a time. The first person may shake hands with other people. The next person may shake hands with other people, not counting the first person again. Continuing like this gives us a total number of handshakes, which is exactly the answer given above. (Retrieved from http://mathworld.wolfram.com/HandshakeProblem.html)

The experience conceived by the student, be it real or imaginarily, when translated into ways of thinking about a situation, is one of the central ingredients of the problem-driven conceptual development.

Lesh and Zawojewski (2007) put it clearly by stating that a problem may be any situation or task where the problem-solver feels the need to find a productive way of thinking about it. Productive ways of thinking do not necessarily mean direct paths between the givens and the goals of the situation; on the contrary, they are the result of seeing the situation in effective ways that may involve several iterations of interpreting, describing and explaining. The proponents of the Models and Modelling Perspective (MMP) have provided a range of evidence that students are able to create conceptual models while devising ways of thinking about a situation. A result of productive thinking is a conceptual model of the situation that includes explicit descriptive or explanatory systems. It is the descriptive and explanatory quality of the thinking that makes it function as a model, an externalisation of the ways in which individuals are actually interpreting a situation and developing means to achieve a problem solution. As Lesh and Doerr (2003a) explain:

> Students produce conceptual tools that include explicit descriptive or explanatory systems that function as models which reveal important aspects about how students are interpreting the problem-solving situations. (Lesh & Doerr, 2003a, p. 9)

This is the fundamental reason why conceptual models usually become expressed in several different representational systems that students may resort to, like images, diagrams, symbols and concrete representational materials, all explicit elements that give visibility to their understandings as, for example, the quantities they think about, the rules they consider, the relationships they establish and so forth.

4.3 Problem-Solving as Expressing Thinking

In the view of problem-solving endorsed here, it is crucial to replace the notion of getting an answer to the problem with the idea of creating an explanation for your answer—a more useful construct that encapsulates both the looked-for answer and the process of finding it. This is an important point since it leads to seeing problem-solving as an activity that includes expressing thinking at its core. Rather than separating the solving stage from the reporting stage, we propose that these are two intimately connected aspects of problem-solving and that such connection is eventually deeper when the use of digital tools is available to support the expression of thinking. Therefore, descriptions, illustrations, explanations and all the material incorporated in the final product are actually the path taken for the product to become a product, as argued by Lesh and Doerr:

> ...descriptions, explanations, and constructions are not simply processes students use on the way to "producing the answer," and, they are not simply postscripts that students give after the "answer" has been produced. They ARE the most important components of the responses that are needed. (Lesh & Doerr, 2003a, p. 3)

Seeing problem-solving in this light is also a consequence of the underlying assumption that it necessarily involves mathematisation—or rather, developing a model and externalising it in some or several ways (Reeuwijk & Wijers, 2003). Thus a mathematical representation, such as an equation or a tree diagram, should not be taken as the reasoning even if it is a key part of the solution process. Instead, it has to be placed within a descriptive story that contains both the particular mathematical representation and the original context of the problem, which is echoed in the following conception of mathematical understanding, as Lester and Kehle (2003) explain:

4.3 Problem-Solving as Expressing Thinking

...a blurring of task, person, mathematical activity, nonmathematical activity, learning, applying what has been learned, and other features of problem solving. (Lester & Kehle, 2003, p. 516)

Thus, when looking at problem-solving, we should be looking primarily at mathematical understandings or, more precisely, mathematical ways of understanding problem situations.

To presuppose that the expression of thinking is an integral part of mathematical problem-solving leads us to consider how such expression takes place today, in digital environments, because expressing mathematical thinking requires a consideration of the means to achieve it.

So, for instance, the photos or the handshake problems referred above reveal a slightly different conceptual structure if the solver chooses to start with the simpler case (two individuals) and increase successively the number of people, like in the following reasoning: two friends take a photo, if another arrives they take two more photos than in the previous case, if another friend arrives they take three more photos than in the previous case, etc. Here, a useful tool to express this thinking would be a spreadsheet that allows the implementation of a relational formula based on a recursive model (Fig. 4.2).

Back in the early eighties, Lesh (1981, p. 254) stated: "research has failed to clarify how technological devices can contribute to the acquisition of the conceptual models that are the most important goals of mathematics instruction; and it has failed to clarify the processes that are needed when mathematical models and technological tools are used to solve problems in real situations". Much more recently Hegedus and Moreno-Armella (2009a) and Santos-Trigo (2004, 2007) have been dealing with this particular issue when addressing problem-solving performance in face of the usage of new digital resources or new representational and communicative infrastructures.

Santos-Trigo has launched a set of questions while arguing that they represent an important direction to current research. Among them, we stress the following: what type of mathematical reasoning do students develop as a result of using computational

Fig. 4.2 A conceptual model of the photos problem expressed in a spreadsheet

	A	B
1	Number of girls	Number of photos
2	2	1
3	3	=B2+A2
4	4	=B3+A3
5	5	=B4+A4
6	6	=B5+A5
7	7	=B6+A6
8	8	=B7+A7
9	9	=B8+A8
10	10	=B9+A9

technology in their problem-solving approaches? What is the students' process of transforming a device, the software, into a mathematical problem-solving tool?

Starting by a case study which documents the solution for a problem by using dynamic geometry software, Santos-Trigo concluded that the systematic use of computer tools can enhance students' problem-solving approaches. He adds that different tools offer different ways to perform and explore mathematical problems and it ends by stating that it will be useful to understand which types of reasoning are developed by students when they use different computer tools.

Hegedus and Moreno-Armella (2009a) documented the usage of representation and communication infrastructures in the mathematics classroom, stressing the new mathematical representational expressivity. According to these researchers, one of the most remarkable aspects within the technological environments is the way expressivity is transformed: there is a new representational expressivity because students can profit from the various software functionalities that permit "natural ways" of expression—metaphors, informal records and deixis (the use of indexical expressions which require a reference to the non-linguistic context of the situation) as well as gestures and movements. In Hegedus and Moreno-Armella's data, actions which seem to be directly connected with the use of technological devices stand out: colouring in strategic ways, using visual gestalts or inserting dots, aiming at underlying the mathematical system. As they claim, "students express themselves in vivid ways, both informally and formally" (2009a, p. 405) which brings new connotations to the act of representing and communicating mathematically in the twenty-first century.

4.3.1 Expository Discourse in Problem-Solving

Problem-solving as a background for mathematisation requires, as said before, a considerable emphasis in the expository and explanatory feature of the developed solutions. Most of the research around problem-solving has approached a large variety of issues: strategic thinking, domain-specific knowledge and metacognitive behaviour are just a few. In the case of the present study, our focus is on the expression of mathematical thinking and, more precisely, on the production of solutions to word problems through the use of technological devices in the context of students' independent computer use.

Some studies have observed the forms of students' written mathematics, sometimes connected with classroom tasks or with assessment assignments. These studies refer to the mathematical writing done with the traditional tools—paper and pencil—and not with another type of tools, such as the computer. Nevertheless, they allow us to have an enlightening vision of some factors that permeate mathematical writing as a way to express thinking in mathematical tasks.

The idea to consider writing as a learning activity was claimed by Shield and Galbraith (1998), supporting themselves on curriculum recommendations from the United States of America and Australia, in the sense that students need to learn how to communicate in a mathematical way and to develop a suitable language to convey what they are learning. In the study they carried out, in three classes of eighth graders, they tried to describe the expository writing of the students prompted by two types of tasks: "write a letter to a friend" and "respond to a student's difficulty". When they define expository writing, they state that this type of writing intends "to describe and explain mathematical ideas" (Shield & Galbraith, 1998, p. 29). While they stress the importance of the prompt to explain, they consider that students' mathematical writing can be seen as an entrance door to check the understanding of particular mathematical ideas held by the students.

One of the indicators of this understanding seems to be the presence of connectors between knowledge units, something that is related to the idea of elaboration. Based on the coding of the written productions done by the students, the results obtained showed that they have, predominantly, an algorithmic style of writing, "with a focus on doing a procedure as an algorithm with few other characteristics of an effective explanation" (Shield & Galbraith, 1998, p. 43).

Shield and Galbraith also concluded that this style can be strongly influenced by the mathematics classroom social context and it can be the reflection of what students consider adequate as mathematical writing, that is to say, with a minimum elaboration. In what concerns the visibility of the students' understanding through the writing they produce, the results were not very conclusive and suggested the need to investigate more about the features of an explanation that can better indicate the students' understanding level.

Ntenza (2006) also calls attention to a growing demand on students to do more writing which includes not only mathematical symbolism but also verbal sentences in ordinary language. The international recommendations have been pointing out this way, when they include a relevant place to the mathematical communication (either as communicational interaction in the classroom or as the development of the competence of using symbols and mathematical representations or else as argumentation and justification in investigative tasks and in mathematical proofing). Ntenza presents a review of the research conducted in the United Kingdom, the United States, Australia and South Africa, showing the following: a very limited use of mathematical writing in mathematics, although such writing becomes more visible when there is the introduction of investigative tasks in the curriculum or the practice of journal writing; most mathematics teachers share a vision that the writing has little to do with mathematics and they do not know for sure the benefits of mathematical writing in students learning; and finally, most of the mathematical writing developed by students is nothing beyond copying and transcribing information.

In his own study of six schools, Ntenza (2006, p. 332) found that "in the majority of schools there were many written pages of mathematics consisting of mathematical symbols rather than some indication of extended mathematical writing". What Ntenza calls mathematical writing (something distinct from symbolic writing) proved to have a limited expression, mainly in terms of a creative use of language.

For the purpose of the present research, the referred results about students' mathematical writing offer some important clues, namely, the importance of a prompt which leads the student to explain his/her thinking and the results he/she obtains, the conception that mathematical writing is much more than symbolic writing and the acknowledgement that expository and explanatory power goes side by side with understanding (or the development of productive ways of thinking about a problem situation).

What we also wish to stress is that mathematical writing can be, as in other school subjects, a form of multimedia writing. This means that communication has, today, several mediational digital tools, very different from the paper and pencil in many ways. It seems to be more adequate to speak of a new mathematical discourse—a digital-mathematical discourse—performed by young people who communicate with digital tools and for whom writing (perhaps typed using a keyboard) continues to be one of the discourse elements, though not the only one, neither necessarily the most evident.

The study developed by Stahl and his collaborators on Virtual Math Teams (Stahl, 2009a) is particularly helpful in offering clues to develop the idea of a digital-mathematical discourse. The Virtual Math Teams project (VMT) consists of one of the many online services offered by the quite well-known Math Forum website, currently accessed by millions of visitors a month. The VMT service has grown out of another service in the Forum, the Problem of the Week (PoW), where challenging mathematical problems are posted and students can send their solutions and receive feedback for improvement. The VMT is a way of working on open-ended problems, in a collaborative mode, with students interacting in groups of peers in mathematical discussion chat rooms. Specific software tools available in the VMT environment allow for maintaining group coordination and mathematical problem-solving, such as the case of the digital whiteboard for graphical representations or the tools to edit mathematical symbols.

Following the research of Stahl and his team, the concept of expository discourse (which they distinguish from exploratory discourse and see as complementary in their data analysis) is an important tool for an analysis of problem-solving as expressing thinking. As Stahl (2009b) describes it, expository discourse is the telling of a story about how a problem was solved, usually providing a sequential account of the essential elements that constitute the problem-solving process.

From this perspective, a large number of signs, considerably propelled by the use of digital tools, become significant as part of an expository discourse: the use of colour, natural language, mathematical language, highlighters, drawings, pictures, photos, icons, diagrams, labels, codes, pre-symbols, symbols, tables, text boxes, outputs of specific software (spreadsheets, dynamic geometry systems, graphing tools) and many others.

Medina, Suthers, and Vatrapu (2009), also reporting on a study about VMT, have considered the question of students' representational group practices and have realised how inscriptions become representations in students' problem-solving attempts. From looking at a group of students engaged in finding a formula to translate a geometrical pattern, the authors highlight how students' inscriptions in the whiteboard guided the group's activity and how they were converted into representational resources with the ability of working as indexical signs to the problem-solvers and to the potential readers.

Other analyses produced within the VMT project focused even more on exploratory discourse, where the mathematical problem was collaboratively formulated, explored and solved in the team discourse itself (Stahl, 2013). In particular, analyses of teams using collaborative GeoGebra showed how technology was enacted through a series of group practices that the team adopted in their problem-solving interactions (Stahl, 2015).

This team discourse is obviously quite different from the one-way speech produced through e-mail by the participants in SUB12 and SUB14. However, commonalities are found in the ways of solving and expressing solutions to problems when making use of the computer (Amado, Carreira, Nobre, & Ponte, 2010; Carreira, 2015; Jacinto & Carreira, 2012; Jacinto, Carreira, & Amado, 2011; Nobre, Amado, & Carreira, 2012). Their expository discourse tends to be rich in inscriptions with strong indexical value. For example, in a problem referring two persons walking towards each other with different speeds, one student's diagram is corroborated with the verbal sentence "the arrows indicate opposite directions of walkers". In fact most of such pieces of information are meaningless without the original context of the problem and outside of the complete story of the problem-solving. But they actually have a profound role not just as a postscript of the problem-solving process but as part of the representational practices students engage in to express their thinking.

4.3.2 Technology Used for Expressing Thinking in Problem-Solving

In discussing the unique contributions of computers to innovative mathematics education, Clements (2000) calls attention to the present scenario of using computers in the classroom, and in this "picture" he shows that the contributions of computers are minor and almost have no role in the developing of mathematical creativity in the following activities: mathematics as problem-solving, mathematics as communication, mathematics as reasoning and making mathematical connections.

It is important to note that all these variants of mathematical activity are relevant in our conception of problem-solving as integrating the expression of thinking. Although problem-centred innovative approaches in the classroom may still be in an insufficient number, the research results about the use of different types of software and of computer tools permit the conclusion that the advantages of using computers can be significant or large, either in problem-solving or in the development of reasoning or still in the learning of content material. Some of the computer potentialities are related to the possibility to emphasise conceptual development and problem-solving, allowing students to build on initial intuitive visual approaches and on informal strategies, promoting a link between formal and visual representations and facilitating the combining of natural language and mathematical language. For example, Clements (2000) refers to the case of using the spreadsheet to generalise patterns in which the language concerning the cells' references and the relations settled by means of formulas becomes a means used by students to express their mathematical reasoning. Thus, it is important to know what happens when students are free to choose the digital tools they have at their disposal to express their mathematical reasoning in solving problems. It is crucial to understand if the tools of general and multipurpose use are recognised and used in an efficient way by the students to develop conceptual models, to take advantage of initial strategies and visual approaches and to establish bridges between the natural language and the symbolic one.

Cook and Ralston (2005), for example, study the role of diagrammatic software in helping young children to create bridges between ideas and visual recording, through story maps or semantic maps. The study involves mainly the use of symbolic, iconic and textual representations, and although it does not pertain to the realm of mathematics education, it highlights important notions related to features of digital expressive technological tools: colour and imagery represent important attributes that impel clarity, attractiveness and immediacy. The students' need to express an idea on the screen, and the search for the best way to accomplish this with the graphic organiser tool was one of the effects of trying to represent visually the thinking developed within the task; as Cook and Ralston (2005, p. 221) say, "The efforts children made to search for images which illustrated their perception, support the view that they were trying to make a visual link".

Other examples of the different nature of expressivity with digital tools, compared to the conventional paper-and-pencil approaches, are offered in different settings by the research developed by Hoyles, Noss and collaborators. Healy and Hoyles (1999, p. 59), for example, reflect on the movement towards more visual and image-related mathematical reasoning activated by new technology's visual possibilities of expression:

> The evolution of technology has opened up new possibilities for visual expression in the process of mathematical reasoning. Images now can be externalized through computer constructions, rendering more explicit previously hidden properties and structures. A visual image can be made open to inspection, an object of reflection, which can serve as a building block in an argument—something concrete rather than transitory and fleeting.

One of their aims was to capture students' responses in terms of the visual and symbolic strategies observed and the connections made between them in tasks referring to pattern generalisation, in different settings—the computer integrated in the task or the computer as a supplementary aid—and with different technological tools (a spreadsheet and a Logo Microworld). In terms of the expressive power of technology, they arrived at a dialectical relation in the computer environments between the means of expression and the expression itself, which the authors claim to be not so apparent in the paper-and-pencil case. In a more recent work, Noss et al. (2009) developed another microworld (called eXpresser) that aimed at offering a medium to think about and express generalisation from visual tilling patterns. Again, one of the main purposes was to create significant means for students to express general

4.3 Problem-Solving as Expressing Thinking

laws of pattern formation with a language that is not the standard algebraic language but has all the conceptual meaning underneath it. With eXpresser, students are invited to create conceptual models about the generalisation of number patterns and to express them at the same time: solving and expressing are fused together through the digital medium, one that is an expressive medium to think with. As Noss et al. (2009) put it:

> The eXpresser is designed to enable students to interact with and create figural patterns of tiles and express relationships that pertain to them, that involve the construction of a language that is not algebra, but could—we hope—be subsequently mapped by the students to algebra. (p. 496)

Expression is seen as something that evolves in combination with reasoning and with the tools that shape reasoning. Therefore, as models evolve, language is expected to evolve, and the expression of thinking is indeed part of models' development.

In connecting their theoretical perspectives to Kaput's fundamental idea of representational infrastructures, Hoyles and Noss (2008) have provided a strong argument to explain how digital technologies in fact shape the mathematics that students may reach, explore and learn. Such explanation relies heavily on the ways that digital media provide new and unconventional means of expression that are tightly connected to the medium in which mathematical ideas are explored. Examples of such digital media can be found in the *autoexpressive* microworlds designed to make mathematical ideas embedded in the tool that simultaneously expresses mathematics in non-standard ways and suggests to the user new forms of expressing mathematical thinking in problem-solving. Noss et al. (2009) clarify:

> This knowledge, or rather the ways in which it is expressed, may not look or sound like standard mathematical discourse: indeed, if the representational system underpinning the tool is non-standard, it follows that the knowledge will be similarly non-standard. This is what the notion of situated abstraction seeks to address: it allows us to recognise and legitimate mathematical expression even when it is remote from (or not represented by) standard mathematical discourse. The notion is particularly salient in computational environments, since it is the nature of interactive, dynamic representations that they encourage expression (and therefore, initially at least, conceptualisation) that diverges from standard mathematics. (p. 92)

The initial conceptualisations, usually very distinct from the standard approaches in mathematics, contrast with the latter in terms of the visual, informal, figurative, intuitive and contextual elements and immediately resonate with the idea of problem-solving and expressing being shaped by computational environments of all sorts (pro-expressive tools as well as autoexpressive tools) that students may resort to in developing their conceptual models of problem situations.

An important point to underline is that digital and multimedia environments place the student in a much more enlarged world of experience when compared with the predominant forms of work done through paper and pencil. This essential feature is well documented by studies which focus on the role of multimedia tools with which students express and present their mathematical thinking (Gadanidis & Geiger, 2010; Stahl, 2009a; Villarreal & Borba, 2010). A common feature to these different environments is their multimedia character, which, in turn, means more and more multi-representational expressivity. The digital-mathematical discourse is not possible just with one language, a single performance system, a unique display support or a sole expository discourse mode. If, since in the 1980s, the introduction of multiple external representations in the learning of mathematics is a matter of study and if the advantages of making students able to move between several representations have been defended, the present multimedia digital technologies emphasise the relevance of the representational practices.

4.4 Multiple External Representations

The work of Ainsworth (1999, 2006) gives an important contribution to understand how the external and multimedia representations can, in different ways, support learning. Ainsworth (1999) reports, for example, that in solving algebra word problems, the students can show six external representations associated to four strategies. Though the multiple external representations may reveal not only strengths but also constraints, we know that the fact that students can use their own external representations allows them to suppress difficulties with a certain strategy or specific representation.

They can compensate for this weakness with the change to another one, more affordable or common. Besides, they have the possibility to combine several representations and explore this combination as a way of not feeling themselves limited by the characteristics of just one.

On the other side, multiple external representations can answer to different details of a task, and each of them can contribute to express different information.

The students' individual characteristics are also considered by Ainsworth (2006). There are several individual features which can influence the way people deal with representations: the familiarity with the representation, to be comfortable with the activity domain, age and cognitive style.

When we look at the ways a great range of students participating in the competitions SUB12 and SUB14 express their mathematical thinking to solve problems, using digital tools, the familiarity with the affordances of different tools to develop certain strategies and representations is, certainly, inevitable. It is clear that for a student who is not much used to a spreadsheet and is more used to tools for text editing and drawing, it will be easier to use the last ones because he/she understands "the format and operators of representation and the relation between the representations and the domain" (Ainsworth, 2006, p. 191).

Nistal, Van Dooren, Clarebout, Elen, and Verschaffel (2009) warn that the choice of the representation(s) depends not only on a good matching with the features of the task to solve but also of the characteristics of the individuals as well as of the context in which the representations are produced. This way, in studies that pay attention to the technological media available for the students "because technology has the potential for broadening the representational horizon", representational fluency is in the foreground of technology-based mathematical activity, problem-solving and student-centred learning (Zbiek, Heid, Blume, & Dick, 2007, p. 1193).

Another important issue is the interaction between external representations and strategies for problem-solving. Strategies and representations are not two separated entities; it is not to develop a strategy and then choose a representation that supports the strategy nor taking a representation and then pick out a strategy. Strategies and representations form the two sides of the same coin. Nistal et al. (2009) argue that research has been more assertive in showing that certain representations trigger the use of certain strategies than consider cross-fertilisation between multiple external representations and the choice of strategies. When they try to join the idea of flexible representational choice to the idea of adaptive strategy choice, they elaborate on a set of factors which influence the choice of a representation; this choice does not always fit the one which would be the best in regard to the hypothetical match between representation and the characteristics of the task. Consequently, the idea that there are representations that would be the best to certain tasks collapses. The context is one of the decisive conditions. If the students feel obliged to use predetermined representations (by curriculum prescription or by the own task determination), their flexibility becomes certainly restricted as the individual particular features and the nature of the task sit on the sidelines.

These authors consider that future research about the role and the features of the context in the choice of representations will improve if researchers pay more attention to the choice of the strategies. The results from our research project have already pointed out that an overall image of representational fluency flows from the expository discourses of the participants in an online competition and such fluency is strongly interlinked with their use of digital media. For example, Jacinto, Amado, and Carreira (2009) looked at how participants in such a competition perceived the role of the technological tools they used during their online participation. The participants valued the opportunity of communicating their reasoning in an inventive way as they could resort to any type of attachments, in particular those they felt more comfortable with or found adequate to the problem itself. They chose mainly the text editor Word, but also the Paint and the Excel, all examples of home digital technology. The use of images was often a result of intentional efforts of expressing their reasoning. Moreover, we noticed consistency between their representations and the way in which they revealed their reasoning. Nobre et al. (2012) also reported on how students dealt with one of the competition problems with the use of a spreadsheet. It was clear that students interpreted the problem in light of their mathematical knowledge and of their knowledge of the digital tool. When the problem was later explored in the classroom with their mathematics teacher, the relationship between the language of the spreadsheet and the algebraic language has shown to be clear to the students.

Computers change the status of visualisation in mathematical activity and bring in new tools to express ideas through visual forms. The media used by students to develop visual representations in problem-solving goes much further than just embellishment. The ways in which students see the problem and express it with digital media supports the statement: "what we see is always shaped by the technologies of intelligence that form part of a given collective of humans-with-media, and what is seen shapes our cognition" (Borba & Villarreal, 2005, p. 99).

The expression of thinking in mathematical problem-solving mediated by students' freely incorporated home computer technologies is therefore pushing the ongoing research to identify the strategies and representations that demonstrate their productive ways of thinking about a problem situation.

In addressing the case of spatial tasks, Nistal et al. (2009) make a reference to the work of Vessey (1991), indicating that such tasks would be facilitated by the use of graphs as those would be more in line with the purpose of dealing with spatial information. However, the data produced within our research project reveal that strategies and representations are variously combined by different students in their digital environments for problem-solving and expressing. This clearly disrupts the idea of "optimal" strategy and "optimal" representation to match the task.

4.5 Humans-with-Media and Co-action with Digital Tools

Borba and Villarreal (2005) emphasise the fact that it is impossible to separate people from technologies, requiring us to think of them as a unit—humans-with-media. From this point of view, humans are constituted by technologies, in the sense that these transform and modify their reasoning but, at the same time, humans are continuously transforming technologies. As a result of this, some dichotomies lose their meaning: subject/technology, technology/knowledge and internal/external representation.

Another significant idea, proposed by Borba and Villarreal, is that our experience with certain media, either present or past, is part of the unit humans-with-media, even if the referred experience may not be available at a certain time. The experience becomes an integral part of a way of cognition, and the computer does not limit itself to assist or help in the accomplishment of certain mathematical procedures but transforms the nature of what is done with its that is to say, it changes the essence of the mathematical activity itself. So, if we think about visualisation, for instance, we should have in mind that what we see is shaped by the technologies of intelligence which are intrinsically part of the unit humans-with-media and, on the other side, what we see shapes also our cognition.

Evidence that the union between technologies and mathematical activity is something deeply transformative can be noticed in the ways that different computer tools shape the processes of mathematical problem-solving (Carreira, 2003, 2009; Carreira et al., 2012; Nobre et al., 2012).

The transformative nature of technologies in the human and cultural procedures is largely discussed by Shaffer and Kaput (1999), who talk about the virtual culture we are dealing with, in which we participate and that is changing us in the most varied ways—a strong concept of mediation that posits the distribution of intelligence among people and objects.

4.5 Humans-with-Media and Co-action with Digital Tools

The technological media are creating new ways of cognitive activity and, with them, a new cognitive culture. For instance, the research reported by Nobre, Amado, Carreira and Ponte (2011) highlights the structure of students' algebraic thinking expressed in a particular representation system shaped by the use of an electronic spreadsheet. It provides a clear display of how students interpreted a problem in light of their mathematical knowledge and of their knowledge of the technological tool. In particular, it highlights the ways in which different modes of creating and exploring rules among cells and columns yields different ways to get the solution. In another study (Nobre et al., 2012), it was possible to observe the prevailing role of the spreadsheet in students' processes of variable identification and understanding of the problem conditions, in their numerical approaches to algebraic models and experimental approaches to solving equations. In the students observed, one could see the phenomenon widely discussed and documented by Borba and Villarreal (2005, p. 98): "the mathematics produced by humans using only paper and pencil will be different from that produced by humans-with-computers".

The same phenomenon was referred by Santos-Trigo (2004) concerning the solution of a problem by 12th graders who had access to Excel, to Cabri-Geometry and to the symbolic calculator, in weekly technology sessions with a problem-solving approach. According to the results of the study, "the use of technology became a powerful tool to explore properties and relationships that did not appear in paper and pencil approaches" (Santos-Trigo, 2004, p. 16).

Changing, reshaping, reorganising and affording are some of the keywords that recent research has been embracing to describe and explain the impact of digital tools in our society. Noss (2001) speaks of the representational transformation as a feature centrally placed at the heart of post-industrial societies and discusses how computational representations are reshaping the nature of mathematical knowledge. There is an increasing awareness of the mediational role of technological-mathematical representations and of how they influence the semiotic registers involved in mathematical activity: mathematical knowledge is inextricably linked to the tools—physical, virtual and cultural—in which it is expressed (Hoyles & Noss, 2009). Accordingly, the role of technology cannot be reduced to the mathematical treatment of a problem within a particular semiotic register or to the conversions between semiotic registers, as Artigue and Bardini (2010) noted in their study. Instead, in describing how students solved a mathematical problem with TI-Nspire CAS, they found that "there is a sophisticated interplay between different instruments belonging to the students' mathematical working space" (Artigue & Bardini, 2010, p. 1179). This is then described as a delicate harmony between the mathematical and the instrumental activity expressed in complex semiotic games that are not simply transitions between representations. Using a different way of addressing the same kind of phenomenon, Hegedus, Donald, and Moreno-Armella (2007) speak about the exploration space in connection to the use of dynamic geometry software and make reference to an emergent distributed intelligence from the co-action between the student and the digital tools: "the complete action-reaction loop exhibited by the co-action of the student and the digital environment, forces cognition to be distributed in the space defined by the agent (the student) and the environment" (p. 1420).

The fundamental point overseeing such observations can be summed up in the idea that individuals and technologies cannot be understood as disjoint sets. The technologies have an impact on the ways to create knowledge which is not consistent with the notion of juxtaposition of persons and technologies or with the perspective of seeing technologies as an addition to the human skills. This is Borba and Villarreal's (2005) overall thesis: to overtake the dichotomous vision between humans and media and proposing a new unit of analysis—the human-computer system.

As we analyse this inter-shaping relationship between the user and the digital tool, we understand how the way of expressing mathematical thinking, that is, the way of transmitting ways to solve problems (students' models) (Santos-Trigo, 2004), is an activity in which the intentionality interacts with the executability of digital representations (Hegedus & Moreno-Armella, 2009b; Moreno-Armella & Hegedus, 2009).

We find a clear example of this type of inter-shaping relationship in the solution proposed by a student to a problem from SUB14 which describes a cat chasing a mouse till grabbing it (Jacinto et al., 2011). The problem states that two steps of the cat make up the same distance as twelve steps of the mouse, and it also mentions that the time spent by the mouse to carry out ten of its short steps is the same as the time spent by the cat to carry out three of its large steps. Besides, when the pursuit starts, the mouse has an advantage of 88 of its short steps on the cat. The problem is to find out how many steps the cat must run to catch the mouse.

Leonor, a 7th grade student, presents the solution in a Word file attached to her e-mail. The answer to the problem includes a description of her way of thinking about the situation.

After having explained, using her own words, all the conditions that must be satisfied to solve the problem, she engages in "creating a diagram" (Fig. 4.3).

The construction of the diagram starts with the drawing of the 88 little steps of the mouse that separate it from the cat, coloured in dark blue small dots: "the first 88 small dark blue dots stand for the 88 little steps advantage that the mouse had over the cat". The dots and lines were created with the software tools that allow inserting and formatting shapes in a page. Instead of drawing the 88 little dots along a single straight line, Leonor arranges the trajectory around a spiral of squared lines and thus she is able to condense the diagram in a considerable small area. Moreover, she signals each set of six small dots by introducing little separators between them. With this she has a representation of the distance covered by six steps of the mouse and equivalently the distance covered by one step of the cat (12 steps of the mouse are equivalent to 2 steps of the cat). By juxtaposing one big dot to each set of six small dots, she devises a way to represent the respective positions of the cat and of the mouse in the run. Moreover, she makes a fundamental use of colours in her drawing. The colours displayed in the little dots and in the big dots have an extremely relevant meaning to expose and explain the solution of the problem. By matching the colour of one set of three big dots with the colour of one set of ten small dots, she is able to display the cat's position in relation to the mouse's position in each moment of the pursuit (Fig. 4.4).

4.5 Humans-with-Media and Co-action with Digital Tools

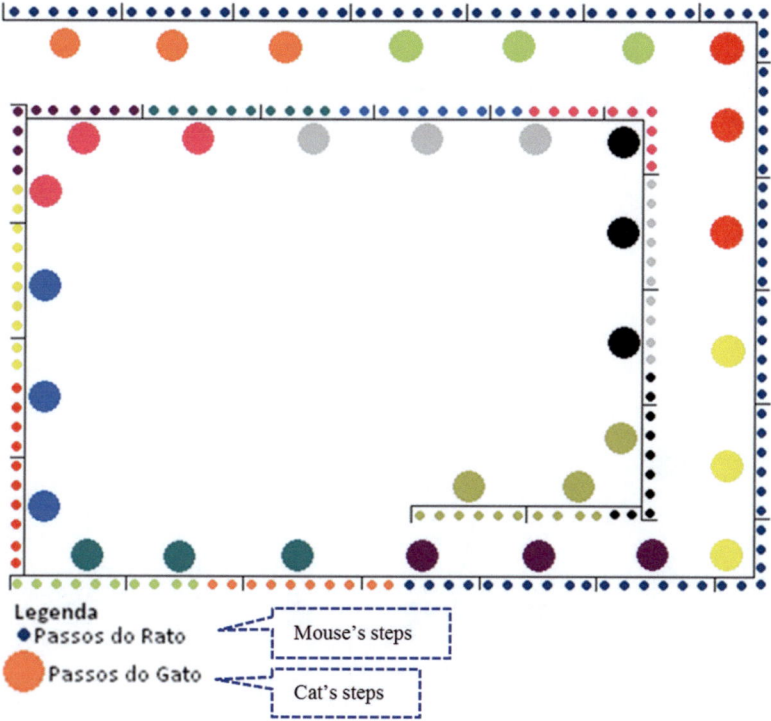

Fig. 4.3 Leonor's diagram in her solution to the problem of the cat and the mouse

The example illustrates not only a simple and clever way of solving the problem, using a concrete visual and iconic approach, but it also reveals an inter-shaping relationship between the solver and the digital tool, where intentionality (depicting the problem situation by means of a diagram) is merged with executability (the iconic representations are associated with manipulation, formatting, customising, setting up, imparting meaning and conveying the solver's thinking). It becomes essential to look at the user and the tool as being a single unit and to understand how that unit operates while solving and expressing problems, namely, by acknowledging the role of visual solution methods and other experimental-like approaches. It elucidates the pairing of individuals' intentionality with the executability of digital representations, in the sense that Hegedus and Moreno-Armella (2009b, p. 398) put it: "Today, besides their traditional spaces, symbols are digital and that means that intentionality becomes blended with the executability of their digital representations".

Another way of looking at the unity between the solver and the digital tool in students' problem-solving and expressing activity concerns the new possibilities found in all sort of technological devices related to imagery and visualisation modes; interactive and manipulative approaches to objects; expressivity in the use of colour, icons and formats; and communication in sound, video, symbols and

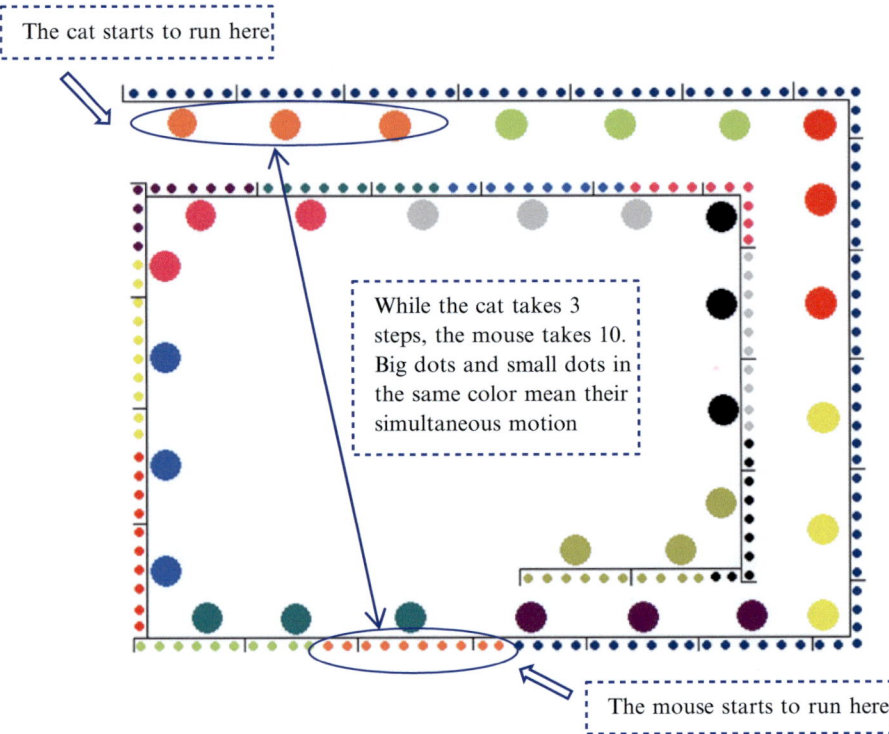

Fig. 4.4 Interpretation of Leonor's thinking embedded in her diagram, emphasising the way in which the cat and the mouse are positioned one relative to the other

pictures. Thinking and communicating are becoming more and more blended under the plasticity of digital artefacts. The evolution from static digital objects to dynamic digital objects has been related to the concept of co-action between the user and the medium where she/he operates (Hegedus & Moreno-Armella, 2009a, 2009b, 2010, 2011; Moreno-Armella & Hegedus, 2009; Moreno-Armella, Hegedus, & Kaput, 2008). The idea of co-action suggests that the individual can be guided or led by a selected digital tool to carry out a certain activity and simultaneously can guide and direct the tool to achieve the purposes at hand. For example, the use of a spreadsheet to solve a problem involving the sum of the first elements of a linear sequence may prompt different ways of thinking; but at the same time, the solution to the problem can be a consequence of the tool guidance to express the thinking as well.

An illustration of these possibilities is given by different approaches to a problem about a staircase posed by SUB12 in 2008–2009 (Fig. 4.5):

The user may decide to create a list with the first even numbers and then use the function SUM, thus guiding his thinking by the possibility of generating linear patterns and using a particular spreadsheet mathematical function to add a certain range of cell values.

4.5 Humans-with-Media and Co-action with Digital Tools

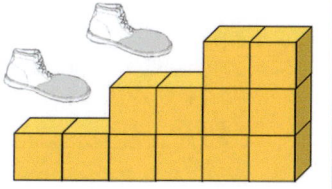

We can build a staircase stacking cubes as shown in the picture. If we want the height of a staircase to be 11 cubes, how many cubes are needed to build that staircase?

Do not forget to explain your problem solving process!

Fig. 4.5 Problem #5 from the 2008–2009 edition of the competition SUB12

| Neste exemplo que nos é dado nós descobrimos que o degrau mais baixo das escadas teria 2 cubos, e vimos também que nos degraus a seguir tinha mais 2 cubos e depois mais 2 cubos e assim sucessivamente.
2+4=6
6+6=12
12+8=20
20+10=30
30+12=42
42+14=56
56+16=72
72+18=90
90+20=110
110+22=132
São necessários 132 cubos para formar essa escada. | In the given example, we have found that the lowest step would have 2 cubes and we have also seen that the following steps would have 2 more cubes and then 2 more cubes and so on.
2+4=6
6+6=12
12+8=20
20+10=30
30+12=42
42+14=56
56+16=72
72+18=90
90+20=110
110+22=132
There will be 132 cubes needed to make the staircase. |

Fig. 4.6 Solution of a SUB12 participant using the successive partial sums

But the recursive working of formulas may also suggest the user that it is possible to generate a column with all the successive partial sums, just by continuously adding the following term to the previous partial sum, as shown in a 6th grader's approach to the problem (Figs. 4.6 and 4.7).

Still another user may take advantage of the rectangular grid to highlight how the staircase is arranged on the grid and finding out an alternative way of rearranging the pattern into a new geometrical figure—a rectangle. And this may result in a new way to generate a formula which calculates the area of the rectangle with lengths n and $n+1$, as in the case of another 6th grader's reasoning (Figs. 4.8 and 4.9).

As Moreno-Armella et al. (2008) describe it, these are some of the immense choice of actions that can be performed with a spreadsheet to accomplish a particular task. These actions are transformed into observable expressions of the solution that are in fact co-actions between the user and the media. What becomes fundamental

	A	B	C
1	Number of steps	Number of bricks	Sum of bricks
2	1	2	=B2
3	2	4	=C2+B3
4	3	6	=C3+B4
5	4	8	=C4+B5
6	5	10	=C5+B6
7	6	12	=C6+B7
8	7	14	=C7+B8
9	8	16	=C8+B9
10	9	18	=C9+B10
11	10	20	=C10+B11
12	11	22	=C11+B12

Fig. 4.7 A possible approach to compute the sum of elements of a linear sequence with a spreadsheet, using a recursive formula

Eu reparei que para saber o número de cubos de cada escada, tenho que multiplicar o número de degraus pelo número que vem a seguir:
escada de 1 degrau – 1 x 2 = 2
escada de 2 degrau – 2 x 3 = 6
escada de 3 degrau – 3 x 4 = 12
escada de 4 degrau – 4 x 5 = 20
escada de 5 degrau – 5 x 6 = 30
E agora só é preciso fazer os cálculos para 11 degraus:
escada de 11 degrau – 11 x 12 = 132
R.: Numa escada de 11 degraus usa-se 132 cubos.

I noticed that to get the number of cubes of each staircase, I have to multiply the number of steps by the number that follows it:
staircase of 1 step – 1 x 2 = 2
staircase of 2 steps – 2 x 3 = 6
staircase of 3 steps – 3 x 4 = 12
staircase of 4 steps – 4 x 5 = 20
staircase of 5 steps – 5 x 6 = 30
And now you just need to do the calculations for 11 steps:
staircase of 11 steps – 11 x 12 = 132
Ans.: On a staircase of 11 steps one uses up 132 cubes.

Fig. 4.8 Solution of a SUB12 participant using the product of two successive integers

is the nature of the feedback offered by the digital environment. Another important aspect of co-action is associated with its strongly framed sociocultural nature. Drawing on the well-known Vygotsky's notion of zone of proximal development, Hegedus and Moreno-Armella (2010) propose the construct of *zone of proximal development of the artefact* to emphasise how the environment dynamically changes with the humans as it opens up new ways of developing knowledge and cognition.

> This type of co-action generates a zone of proximal development (ZPDA) for the artifact that will be realized (in human activity) as long as the artifacts are stable and full of visibility in the cultural/technological medium. (Hegedus & Moreno-Armella, 2010, p. 31)

4.5 Humans-with-Media and Co-action with Digital Tools

	A	B	C	D	E	F	G	H	I	J	K	L	M	N	O
1	Number of steps	Number of cubes													
2	1	=A2*(A2+1)													
3	2	=A3*(A3+1)													
4	3	=A4*(A4+1)		5											
5	4	=A5*(A5+1)		4											
6	5	=A6*(A6+1)		3											
7	6	=A7*(A7+1)		2											
8	7	=A8*(A8+1)		1											
9	8	=A9*(A9+1)			1	2	3	4	5	6	7	8	9	10	
10	9	=A10*(A10+1)													
11	10	=A11*(A11+1)													
12	11	=A12*(A12+1)													
13															

Fig. 4.9 The rectangular grid of the spreadsheet as a means to visualise a rearrangement of the sum of the first elements of a linear sequence and its translation into the formula for the area of a rectangle

Co-action has another central element that stands on intentionality. Both the human and the media are carrying a load of intentionality. The hammer is intended to hammer, and it can be used in a number of different ways; the user who wants to drive a nail on the wall has an intention that adjusts to the intentionality of the hammer, but the tool does not react to the user's intentionality and therefore it is just a question of a one-direction matching. With digital and particularly with dynamical tools, a large part of the intentionality of the tool is to react to the user, thus becoming a plastic environment, as opposed to the inert hammer that will not react back to the action of the user. The difference is that the user and the environment both become actors and re-actors. They act and re-act on each other and it is such co-action that constitutes a distinctive feature of the unity between user and tool. The symbiosis of humans-with-media is expressed and becomes visible through the development of co-actions.

Gill and Borchers (2003) also consider the notion of knowledge in co-action and relate it to social intelligence. They argue that tools and technologies can be seen as dynamic representations of knowledge which become invisible, like an extension of the self, when one understands how to use it. This entails seeing cognition as a dynamic system that co-evolves through the interaction between mind, body and environment. The development of knowledge in co-action means that the person is able to understand and react to the representations of his/her actions in a way that is suitable and useful to the pursued goal. Co-action does not occur solely in front of a computer or other technology in a solitary activity; it is happening with the environment where other individuals are acting too, where an important role of the tool is to enable humans to engage with each other, to communicate and to create knowledge.

> A challenge for designing mediating interfaces is for them to afford us our human skills of engaging with each other, communicating information and forming knowledge. (Gill & Borchers, 2003, p. 54)

The idea that tools and humans evolve together is essential to grasp the importance of co-action in a digitally permeated environment of youngsters solving mathematical problems, as it is the case of the empirical context of the present research. It hinges on the perspective of seeing cultural tools becoming cognitive tools in the thinking and expressing of many of our young students. Rejecting the separation between humans and media, when looking at the ways in which students approach mathematical problems with digital tools, also means that students and tools are agents performing knowledge in co-action. There is knowledge embedded in the digital environment, which the user can explore, interpret, understand and use in co-action. The product is always an effect of a co-action, and it is also an exposure of the tool's built-in mathematical, spatial, computational, graphical and other kinds of knowledge and of the solver's knowledge. The resultant knowledge is a fusion of both, and it becomes a unit, rather than an aggregate of both or an extension of any of them.

> Humans have been saturating the environment through their artifact-mediated-activities. The outcome is a medium that reflects humans because the environment is re-created through these activities. But humans are not cognitively immune through this process: they are subjected to ever-deeper enculturation. (Moreno-Armella & Hegedus, 2009, p. 518)

Another way of placing this perspective is to accept that the environment has now more participants: digital and human (Moreno-Armella et al., 2008). The digital tools are now an infrastructural part of a learning environment (Hegedus & Moreno-Armella, 2010). They embody knowledge that can be executed through co-action. Such knowledge may be conceived as a sort of *invisible hand* or *invisible intelligence* that projects intentionality. To understand such intelligence (in many cases, this means to understand a set of rules or constraints of the tool together with mathematical properties, symbols and objects that are crystallised in the tool's setup) entails deliberately acting and re-acting on it with the conscious idea that it also acts and re-acts on oneself by means of intentionality. The emergent knowledge from a digital medium is inextricably linked to the mediating artefact, which is all but a neutral one.

4.6 An Outlook

Bearing in mind the digital communication context in which the mathematical problem-solving competitions are anchored, this chapter has brought forth the theoretical perspectives that are considered particularly relevant for describing, analysing and understanding the phenomenon under study, that of students solving mathematical problems with the digital technologies of their choice. Based on a critical review of the literature, the most significant conceptual tools that frame the

4.6 An Outlook

distinctive features of that problem-solving activity are intertwined, leading to a new well-suited approach for the interpretation of how young participants are able to find the solution to a word problem and express their reasoning using digital tools.

Solving mathematical problems within the online competitions' context is firstly equated as a mathematisation activity, where students develop a productive way of thinking about the situation which, in turn, opens the way to a conceptual model. These conceptual models, yielding the solutions to the problems, often become expressed in several different representational systems that, due to discourse elements that the digital tools allow to incorporate, are instances of a new communication system, named digital-mathematical discourse. Problem-solving is, thus, conceptualised as a concurrent process of mathematisation and of expressing mathematical thinking or, in a condensed manner, as a solving and expressing activity. The subjects that are responsible for carrying out such tasks have to be regarded in light of their symbiotic relationship with the technological tools chosen for solving and expressing and a closer analysis of that activity can be reached if one considers the students and the tools as both being agents performing knowledge in cycles of co-action.

We will next present several cases documenting the new approaches of young people who develop powerful conceptual models and express their mathematical thinking through a close relationship with digital technologies. It is that close relationship that encapsulates much of what matters for research in mathematics education: to learn more about how students use their digital skills, no matter their level of proficiency, for seizing productive ways of thinking and engaging in digital-mathematical discourses around problem-solving.

We again stress the idea of positing a unity between solving a problem and expressing mathematical thinking. Awareness of this unity has an important effect on the way we view the production of solutions to problems in the online mathematical competitions SUB12 and SUB14: the digital product that is presented as an answer to a problem accounts for such solving and expressing and is not seen as the student's translation of a strategy or model into a digital medium. Therefore, the attention given to the form of using a digital tool, however simple and commonplace or more sophisticated, derives from looking at technology as more than a presentation device. It entails also endorsing another unity: that of the subjects and the tools being co-agents in the construction of conceptual models, ranging in levels of sophistication, abstraction and generality and integrating a multiplicity of external representations in a close relation to the potentialities that each individual draws from them.

The following three chapters are characterised by a strong emphasis on the analysis of empirical data gathered in the context of the online competitions. They intend to highlight the many ways in which students can achieve the solution to a problem and communicate it mathematically with the technological tools they have at their disposal, either in their home environment or in their mathematics classroom. Those data reveal, moreover, that problems are open to multiple alternative conceptual models which clearly intersect with the affordances of the technologies used.

The main theoretical constructs presented so far are exploited onwards in the subsequent chapters. At the same time, the use of this framework, almost as if it were a "tool box", required a further development of some of those concepts and even the introduction of new theoretical perspectives accordingly to the research questions and the specific features of the data considered in each analysis.

References

Ainsworth, S. (1999). The functions of multiple representations. *Computers and Education, 33*, 131–152.

Ainsworth, S. (2006). DeFT: A conceptual framework for considering learning with multiple representations. *Learning and Instruction, 16*, 183–198.

Amado, N., Carreira, S., Nobre, S., & Ponte, J. P. (2010). Representations in solving a word problem: The informal development of formal methods. In M. F. Pinto & T. F. Kawasaki (Eds.), *Proceedings of the 34th Conference of the International Group for the Psychology of Mathematics Education* (pp. 137–144). Belo Horizonte, Brazil: PME.

Artigue, M. & Bardini, C. (2010). New didactical phenomena prompted by TI-nspire specificities: The mathematical component of the instrumentation process. In V. Durand-Guerrier, S. Soury-Lavergne, & F. Arzarello (Eds.), *Proceedings of CERME 6* (pp. 1171–1180). Lyon, France: INRP.

Borba, M., & Villarreal, M. (2005). *Humans-with-media and the reorganization of mathematical thinking: Information and communication technologies, modeling, experimentation and visualization*. New York, NY: Springer.

Carreira, S. (2003). Problem solving with technology: How it changes students' mathematical activity. In T. Triandafillidis & K. Hatzikiriakou (Eds.), *Proceedings of the 6th International Conference on Technology in Mathematics Teaching (ICTMT 6)* (pp. 67–74). Athens, Greece: New Technologies Publications.

Carreira, S. (2009). Matemática e tecnologias: Ao encontro dos "nativos digitais" com os "manipulativos virtuais". *Quadrante, 18*(1 & 2), 53–86.

Carreira, S. (2015). Mathematical problem solving beyond school: Digital tools and students mathematical representations. In J. S. Cho (Ed.), *Selected Regular Lectures from the 12th International Congress on Mathematical Education* (pp. 93–113). New York: Springer.

Carreira, S., Amado, N., Ferreira, R. A., & Silva, J. (2012). A web-based mathematical problem solving competition in Portugal: Strategies and approaches. In *Pre-Proceedings of the 12th International Congress on Mathematical Education (ICME 12) – Topic Study 34* (pp. 3098–3107). Seoul, South Chorea: ICMI.

Clements, D. (2000). From exercises and tasks to problems and projects: Unique contributions of computers to innovative mathematics education. *Journal of Mathematical Behavior, 19*, 9–47.

Cook, D., & Ralston, J. (2005). Building the cognitive bridge: Children, information technology and thinking. *Education and Information Technologies, 10*(3), 207–223.

Doorman, L., & Gravemeijer, K. (2009). Emergent modeling: Discrete graphs to support the understanding of change and velocity. *ZDM: International Journal on Mathematics Education, 41*, 199–211.

English, L., & Sriraman, B. (2010). Problem solving for the 21st century. In B. Sriraman & L. English (Eds.), *Theories of mathematics education: Seeking new frontiers* (pp. 263–295). New York, NY: Springer.

Francisco, J. M., & Maher, C. A. (2005). Conditions for promoting reasoning in problem solving: Insights from a longitudinal study. *Journal of Mathematical Behavior, 24*, 361–372.

Gadanidis, G., & Geiger, V. (2010). A social perspective on technology-enhanced mathematical learning: From collaboration to performance. *ZDM: International Journal on Mathematics Education, 42*, 91–104.

References

Gill, S., & Borchers, J. (2003). Knowledge in co-action: Social intelligence in collaborative design activity. *AI and Society, 17*, 322–339. doi:10.1007/s00146-003-0286-6.
Gravemeijer, K. (1994). *Developing realistic mathematics education*. Utrecht, The Netherlands: Cdß Press.
Gravemeijer, K. (1997). Solving word problems: A case of modelling? *Learning and Instruction, 7*(4), 389–397.
Gravemeijer, K. (2002). Preamble: From models to modeling. In K. Gravemeijer, R. Lehrer, B. Van Oers, & L. Verschaffel (Eds.), *Symbolizing, modeling and tool use in mathematics education* (pp. 7–22). Dordrecht, The Netherlands: Kluwer.
Gravemeijer, K. (2007). Emergent modelling as a precursor to mathematical modelling. In W. Blum, P. L. Galbraith, H.-W. Henn, & M. Niss (Eds.), *Modelling and applications in mathematics education. The 14th ICMI study* (pp. 137–144). New York, NY: Springer.
Gravemeijer, K., Lehrer, R., van Oers, B., & Verschaffel, L. (Eds.). (2002). *Symbolizing, modeling and tool use in mathematics education*. Dordrecht, The Netherlands: Kluwer.
Healy, L., & Hoyles, C. (1999). Visual and symbolic reasoning in mathematics: Making connections with computers? *Mathematical Thinking and Learning, 1*(1), 59–84.
Hegedus, S., Donald, S., & Moreno-Armella, L. (2007). Technology that mediates and participates in mathematical cognition. In D. Pitta-Panzi & G. Philippou (Eds.), *Proceedings of CERME 5* (pp. 1419–1428). University of Cyprus.
Hegedus, S., & Moreno-Armella, L. (2009a). Intersecting representation and communication infrastructures. *ZDM: International Journal on Mathematics Education, 41*, 399–412.
Hegedus, S., & Moreno-Armella, L. (2009b). Introduction: The transformative nature of "dynamic" educational technology. *ZDM: International Journal on Mathematics Education, 41*, 397–398.
Hegedus, S., & Moreno-Armella, L. (2010). Accommodating the instrumental genesis framework within dynamic technological environments. *For the Learning of Mathematics, 30*(19), 26–31.
Hegedus, S., & Moreno-Armella, L. (2011). The emergence of mathematical structures. *Educational Studies in Mathematics, 77*, 369–388.
Hoyles, C., & Noss, R. (2008). Next steps in implementing Kaput's research programme. *Educational Studies in Mathematics, 68*, 85–97.
Hoyles, C., & Noss, R. (2009). The technological mediation of mathematics and its learning. *Human Development, 52*(2), 129–147.
Jacinto, H., Amado, N., & Carreira, S. (2009). Internet and mathematical activity within the frame of "Sub14". In V. Durand-Guerrier, S. Soury-Lavergne & F. Arzarello (Eds.), *Proceedings of the Sixth Congress of the European Society for Research in Mathematics Education* (pp. 1221–1230). Lyon, France: INRP.
Jacinto, H., & Carreira, S. (2012). Problem solving in and beyond the classroom: perspectives and products from participants in a web-based mathematical competition. In *Pre-Proceedings of the 12th International Congress on Mathematical Education (ICME 12) – Topic Study 15* (pp. 2933–2942). Seoul, South Corea: ICMI.
Jacinto, H., Carreira, S., & Amado, N. (2011). Home technologies: How do they shape beyond-school mathematical problem solving activity? In M. Joubert, A. Clark-Wilson, & M. McCabe (Eds.), *Proceedings of the 10th International Conference on Technology in Mathematics Teaching (ICTMT10)* (pp. 160–165). Portsmouth, UK: University of Portsmouth.
Lesh, R. (1981). Applied mathematical problem solving. *Educational Studies in Mathematics, 12*, 235–364.
Lesh, R. (2000). Beyond constructivism: Identifying mathematical abilities that are most needed for success beyond school in an age of information. *Mathematics Education Research Journal, 12*(3), 177–195.
Lesh, R., & Doerr, H. (2003a). Foundations of a models and modeling perspective on mathematics teaching, learning, and problem solving. In R. Lesh & H. Doerr (Eds.), *Beyond constructivism – models and modeling perspectives on mathematics problem solving, learning, and teaching* (pp. 3–33). Mahwah, NJ: Erlbaum Associates.
Lesh, R., & Doerr, H. (Eds.). (2003b). *Beyond constructivism – models and modeling perspectives on mathematics problem solving, learning, and teaching*. Mahwah, NJ: Erlbaum Associates.

Lesh, R., & Harel, G. (2003). Problem solving, modeling, and local conceptual development. *Mathematical Thinking and Learning, 5*(2 & 3), 157–189.

Lesh, R., & Zawojewski, J. (2007). Problem solving and modeling. In F. Lester (Ed.), *Second handbook of research on mathematics teaching and learning* (pp. 763–804). Charlotte, NC: Information Age Publishing and National Council of Teachers of Mathematics.

Lester, F., & Kehle, P. (2003). From problem solving to modeling: The evolution of thinking about research on complex mathematical activity. In R. Lesh & H. Doerr (Eds.), *Beyond constructivism – models and modeling perspectives on mathematics problem solving, learning, and teaching* (pp. 501–517). Mahwah, NJ: Erlbaum Associates.

Medina, R., Suthers, D., & Vatrapu, R. (2009). Representational practices in VMT. In G. Stahl (Ed.), *Studying virtual math teams* (pp. 185–205). New York, NY: Springer.

Moreno-Armella, L., & Hegedus, S. (2009). Co-action with digital technologies. *ZDM: International Journal on Mathematics Education, 41*, 505–519.

Moreno-Armella, L., Hegedus, S., & Kaput, J. (2008). From static to dynamic mathematics: Historical and representational perspectives. *Educational Studies in Mathematics, 68*, 99–111.

Nistal, A., Van Dooren, W., Clarebout, G., Elen, J., & Verschaffel, L. (2009). Conceptualising, investigating and stimulating representational flexibility in mathematical problem solving and learning: A critical review. *ZDM: International Journal on Mathematics Education, 41*, 627–636.

Nobre, S., Amado, N., Carreira, S., & Ponte, J. (2011). Algebraic thinking of grade 8 students in solving word problems with a spreadsheet. In M. Pytlak, E. Swoboda & T. Rowland (Eds.), *Proceedings of the 7th Congress of the European Society for Research in Mathematics Education* (pp. 521–531). Poland: ERME, University of Rzeszów.

Nobre, S., Amado, N., & Carreira, S. (2012). Solving a contextual problem with the spreadsheet as an environment for algebraic thinking development. *Teaching Mathematics and Its Applications, 31*(1), 11–19. doi:10.1093/teamat/hrr026.

Noss, R. (2001). For a learnable mathematics in the digital cultures. *Educational Studies in Mathematics, 48*(1), 21–46.

Noss, R., Hoyles, C., Mavrikis, M., Geraniou, E., Gutierrez-Santos, S., & Pearce, D. (2009). Broadening the sense of "dynamic": A microworld to support students' mathematical generalisation. *ZDM: International Journal on Mathematics Education, 41*, 493–503.

Ntenza, S. P. (2006). Investigating forms of children's writing in grade 7 mathematics classrooms. *Educational Studies in Mathematics, 61*, 321–345.

Reeuwijk, M., & Wijers, M. (2003). Explanations why? The role of explanations in answers to (assessment) problems. In R. Lesh & H. Doerr (Eds.), *Beyond constructivism – models and modeling perspectives on mathematics problem solving, learning, and teaching* (pp. 191–202). Mahwah, NJ: Erlbaum Associates.

Santos-Trigo, M. (2004). The role of technology in students' conceptual constructions in a sample case of problem solving. *Focus on Learning Problems in Mathematics, 26*(2), 1–17.

Santos-Trigo, M. (2007). Mathematical problem solving: An evolving research and practice domain. *ZDM: International Journal on Mathematics Education, 39*, 523–536.

Shaffer, D., & Kaput, J. (1999). Mathematics and virtual culture: An evolutionary perspective on technology and mathematics education. *Educational Studies in Mathematics, 37*, 97–119.

Shield, M., & Galbraith, P. (1998). The analysis of student expository writing in mathematics. *Educational Studies in Mathematics, 36*, 29–52.

Stahl, G. (Ed.). (2009a). *Studying virtual math teams*. New York, NY: Springer.

Stahl, G. (2009b). Interactional methods and social practices in VMT. In G. Stahl (Ed.), *Studying virtual math teams* (pp. 41–55). New York, NY: Springer.

Stahl, G. (2013). *Translating Euclid: Designing a human-centered mathematics*. San Rafael, CA: Morgan & Claypool Publishers.

Stahl, G. (2015). *Constructing triangles together: The development of mathematical group cognition*. Retrieved from http://gerrystahl.net/elibrary/analysis/analysis.pdf.

Vessey, I. (1991). Cognitive fit: A theory-based analysis of the graph versus tables literature. *Decision Sciences, 22*, 219–240.

Villarreal, M., & Borba, M. (2010). Collectives of humans-with-media in mathematics education: Notebooks, blackboards, calculators, computers… and notebooks throughout 100 years of ICMI. *ZDM: The International Journal on Mathematics Education, 42*(1), 49–62.

Zawojewski, J., & Lesh, R. (2003). A models and modeling perspective on problem solving. In R. Lesh & H. Doerr (Eds.), *Beyond constructivism – models and modeling perspectives on mathematics problem solving, learning, and teaching* (pp. 317–336). Mahwah, NJ: Erlbaum Associates.

Zbiek, R. M., Heid, M. K., Blume, G., & Dick, T. P. (2007). Research on technology in mathematics education: The perspective of constructs. In F. Lester (Ed.), *Second handbook of research on mathematics teaching and learning* (pp. 1169–1207). Charlotte, NC: Information Age Publishing and National Council of Teachers of Mathematics.

Chapter 5
Digitally Expressing Conceptual Models of Geometrical Invariance

Abstract This chapter develops around two fundamental ideas, namely, that (1) the perception of the affordances of a certain digital tool is essential to solving mathematical problems with that particular technology and that (2) the activity thus undertaken stimulates different mathematising processes which, in turn, result in different conceptual models. Looking thoroughly, from an interpretative perspective, at four solutions to a particular geometry problem from participants who decided to use dynamic geometry software at some point of their solving activity, our main purpose is to illustrate the ways in which the same tool affords different approaches to the problem in terms of the conceptual models developed for studying and justifying the invariance of the area of a triangle. Their different ways of dealing with the tool and with mathematical knowledge are interpreted as instances of students-with-media engaged in a "solving-with-dynamic-geometry-software" activity, enclosing a range of procedures brought forth by the symbioses between the affordances of the dynamic geometry software and the youngsters' aptitudes. The analysis shows that different people solving the same problem with the same digital media and recognising a relatively similar set of affordances of the tool produce different digital solutions, but they also generate qualitatively different conceptual models, in this case, for the invariance of the area.

Keywords Affordances • Humans-with-media • Mathematising • Conceptual models • Geometrical thinking • Dynamic geometry • GeoGebra • Invariance

5.1 Main Theoretical Ideas

Nowadays, we live in an e-permeated society (Martin & Grudziecki, 2006) which is also mathematised at the highest level (Noss, 2001). The impact of digital technologies is felt mainly on the ways of representing knowledge. Noss (2001) suggests that the capacity to deal with these new representation forms, which are changing the very nature of mathematical knowledge, may include analysis and critical skills, which are needed to deal critically with digital representations or to develop mathematical modelling of different daily situations.

Adding to the recognition of the more and more meaningful role that mathematics plays in society, there is an increasing concern with the development of technological tools shown to be masking or hiding direct access to mathematical knowledge. One of the effects of this technological revolution consists, precisely, in the fact that technology is less and less open to inspection and understanding: it is no longer possible to find out how a digital watch works just by opening it. Technologies that, some time ago, could have been easily looked into by curious children are nowadays, in a way, impervious to them (Noss, 2001).

5.1.1 Perceiving Affordances of Digital Tools

The term affordance has been used to define the set of special characteristics associated with a certain technological tool, which invite the tool user to execute an action with it (Artigue, 2007; Noss, 2001). The concept has been used according to very different perspectives, mainly to deal with questions related to the person-machine interactions, but it is also found in the literature referring to the usage of technologies in mathematical education (Brown, Stillman, & Herbert, 2004; Drijvers, Godino, Font, & Trouche, 2013; Trouche, Drijvers, Gueudet, & Sacristan, 2013).

The word "affordance" was coined by Gibson when he tried to understand what motivates human behaviour, adopting an ecological approach which determines that the environment's perception necessarily motivates any type of action (Martinovic, Freiman, & Karadag, 2013). Gibson (1986), in opposition to the cognitive perspectives of the time (Greeno, 1994), used key ideas derived from the gestalt psychological approach to explain that the way we perceive the meaning or the value of something is as immediate as the perceiving of its colour: "each thing says what it is… a fruit says 'Eat me'; water says 'Drink me'; thunder says 'Fear me'" he states, quoting Koffka (Gibson, 1986, p. 138). Nevertheless, and to the psychologist Koffka, this required character is liable to undergo changes if the observer needs change. The valence of the object, almost as a physiognomy trait, is recognised by the observer's experienced needs (e.g. water invites to drink if the observer is thirsty). Although Gibson recognises that his concept of affordance derives from others, such as that of valence, invitation or demand, he stresses that it contains a crucial difference since the affordance of something does not change according to the observer needs as he "may or may not perceive or attend to the affordance, according to his needs, but the affordance, being invariant, is always there to be perceived" (p. 139). Brown et al. (2004) illustrate this feature of the affordance invariance: if we observe a rectangular table from different angles, the table does not change, but what we perceive of the table changes. This set of invariabilities of an object, which is understood by the observer, represents the information that specifies the affordance of the object.

Gibson (1977) thinks of an affordance as a previous condition for enabling an activity; that is to say, it defines a set of possible actions between the object and the subject, although the existence of affordances does not always determine that an activity takes place. He then assumes that the affordances are interactions between the subject and the environment, in the same way as Greeno (1994) who names

5.1 Main Theoretical Ideas

them "agent-situation interactions" (p. 338). The conditions that make possible the interaction between an agent and another system include, necessarily, some properties of the agent and some properties of that system. Greeno (1994) also stresses that if the word affordance refers to something that exists in the system which contributes to the type of interaction that occurs, then it becomes necessary to use another expression that designates something that exists in the agent that also contributes to the same situation—suggesting the words "ability" or "aptitude" (p. 338). It is equally interesting and useful to recognise the impossibility of separation between the system's affordance and the agent's aptitude, i.e. the affordance and the aptitude are not specifiable in the absence of one of them. Later, we return to this question.

Meanwhile, we illustrate the sense to be given to the concept of affordance based on the experience which has permitted us to look closely into the ways the participants in the competitions SUB12 and SUB14 use digital tools to their advantage.

The organising committee emphasises the freedom to choose the tools the participants may use to solve the problems and present their strategies. Their option is, normally, prompted by the degree of familiarity they have with certain tools, that is to say, to their own perception of how a particular tool will contribute to obtain the solution or express, as faithfully as possible, the process of finding it.

The solution that Marco (one of the participants in SUB14) sent as an answer to the problem "United and Cropped" (Fig. 5.1), by using a dynamic geometry program, illustrates this issue.

Marco decided to use GeoGebra in his approach to the problem (see Fig. 5.2). A short analysis enables us to state that Marco recognises a range of affordances in this computational tool. At first, he seems to have recognised great potentialities in the combination of two affordances of the graphics view—the rectangular axes and the square grid—that provide a visual support on which he will build the sequence of squares mentioned in the problem statement. This construction starts by marking the vertices of the eight squares, using the tool "new point". Each point is marked on a grid, considering its coordinates and the dimensions of the sides of each square.

After marking the 25 points, he draws the segments that represent the sides of the various squares using the tool "segment between two points". To conclude the

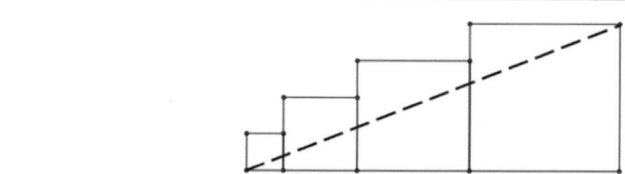

Consider a sequence of squares of sides 1, 2, 3 ... 4 cm, arranged so as to be joined to each other, as shown in the figure. Once together, we cut up all the squares along a line from the lower left corner of the smaller square to the upper right corner of the larger square. What is the area above the cutoff line if the sequence has 8 squares?

Do not forget to explain your problem solving process!

Fig. 5.1 Problem #5 from the SUB14 competition (edition 2011–2012)

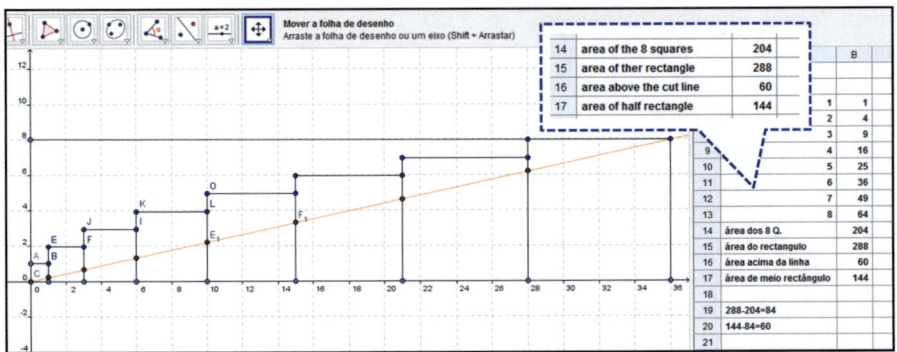

Fig. 5.2 Solution sent by Marco (8th grader)

representation of the situation, he draws a ray starting on the origin and passing through the right superior vertex of the larger square. In the properties of the object, he changes the colour of the ray to orange, and immediately, he determines its intersections with the sides of the squares.

In the following phase, Marco considered it relevant to show the spreadsheet view where he registered the length of the sides and the area of the corresponding square (columns A and B).

Later, he built the upper side of the rectangle, bounding the sequence of squares, and inputted its area on the spreadsheet. He also filled in the intermediate values, which allowed him to determine the answer to the problem: the area occupied by the eight squares, the area of the bounding rectangle and its half as well as the area outside the sequence of the squares and inside the bounding rectangle. The solution is given by the difference between the area of the semi-rectangle, which is above the cutting line and the previously determined measure, which he uses to fill the cell B19.

Hence, Marco recognises a certain convenience in using GeoGebra to solve this problem; that is to say, the affordances he can perceive allow him the activity of organising a strategy and getting a solution. Besides the affordances that are directly related to the construction, this solution presents two others with special relevance in the youngster's activity: on one hand, the grid promotes a "materialisation", almost immediate, of the squares' vertices; on the other hand, the spreadsheet allows him to record every step of his strategy, which includes his reasoning and the procedures taken.

The intentional choice of this program is based on an explicit knowledge of the software potentialities, of its language and tools and of the user's skills, that is, of the things that the participant is able to do with GeoGebra and within GeoGebra to solve this problem.

In spite of the fact that Gibson initially suggested that the affordances are resources that the environment offers to the subject who has the capacity to perceive and use them, several authors in the field of ecological psychology have developed this theory further. More recently, Chemero (2001, 2003) proposes that the affordances neither are properties of the environment nor are found in the environment. Actually, he argues that perceiving the affordances of an environment is placing features in the environment. By admitting that to locate them physically may not be an easy task, he considers that they are "perfectly real and perfectly perceivable" (Chemero, 2003, p. 191), and he adds that "perceiving affordances is placing features, seeing that the situation allows a certain activity" (p. 187).

Stoffregen (2003) recognises that the affordances are central to an ecological approach of the perception and of the action and proposes a formal definition of the concept with the assumption that affordances are properties that arise from the animal-environment system. However he alerts:

> For living things, the conjunction of particular properties of the animal with particular properties of the environment does not lead to the involuntary actualization of the action afforded. Affordances are what one *can* do, not what one *must* do (Stoffregen, 2003, p. 119, italics in the original).

This relation, between environment and subject, objectified in the notion of "system" was already present, though in an embryonic stage, in Gibson's (1986) perspective. Besides, Gibson recognised that the concept of affordance allows the overcoming of some issues related to the dichotomist nature which still persisted in some psychological perspectives.

> An affordance cuts across the dichotomy of subjective-objective and helps us to understand its inadequacy. It is equally a fact of the environment and a fact of behavior. It is both physical and psychical, yet neither. An affordance points both ways, to the environment and to the observer (Gibson, 1986, p. 129).

This claim supports the perspective of several authors who maintain that the affordances, not only being a set of characteristics of the object, arise from the interactions between the agent and the object (Chemero, 2003; Stoffregen, 2003).

The idea of indivisibility between the subject and the context, a salient feature of the ecological perspectives in rejecting the duality person/environment, has appeared in other fields of investigation. Various key points found in Gibson's theory have been addressed in the literature concerning different aspects of mathematics education, but especially in what concerns the use and impact of technological tools on the production and development of mathematical knowledge.

5.1.2 The Indivisibility Between the Subject and the Context

Assuming a similar perspective to that of Gibson and informed by Lévy (1990), Tikhomirov (1981) and Borba and Villarreal (2005) argue that technological tools neither replace nor merely complement human beings in their cognitive activities. The authors claim that the procedures mediated by technologies lead to a reorganisation of the human mind and their standpoint in that knowledge itself results from a symbiosis between humans and the technology they act with. This close relationship generates a new entity which they name humans-with-media (Borba & Villarreal, 2005). Accordingly, this metaphor aims to emphasise how mathematical thinking is reorganised in the presence of the technologies of intelligence or, in other words, to point out the transformations which may occur in the practical activity of this new entity.

Borba and Villarreal (2005) use two fundamental ideas to support the notion of humans-with-media: while they consider the social and collective nature of cognition, they also assume that cognition includes the tools that mediate the production of knowledge. These ideas sum up a central feature of their theory: the media are considered a constitutive part of the subject and cannot be seen as merely aids or complements of the activity.

This entails a sociocultural perspective of the human mind, in a close sense to that proposed by Wertsch (1991) when he assumed that all the "action is mediated and it cannot be separated from the milieu in which it is carried out" (p. 18). In fact, by trying to understand the human mind, Wertsch (1991) proposes the unit "person(s)-acting-with-mediational-means" as the agent who undertakes an action, and he explains that, from this point of view, "any tendency to focus exclusively on the action, the person(s), or the mediational means in isolation is misleading" (p. 119). In a certain way, he emphasises the action, and he seems to assume that the affordances do not depend exclusively on the artefact but mainly on the way the subject uses it during the activity, i.e. aptitude.

It is possible, therefore, to accept that the technological tools used to communicate and to produce or represent mathematical ideas influence the type of mathematics, the mathematical thinking and the mathematical representations which result from that activity. It is in this sense that Borba and Villarreal (2005) recognise that the notion of humans-with-media can be articulated with the theory of multiple representations but only when we agree that graphical representations drawn with paper-and-pencil are qualitatively different from those produced on the computer with some given software. This way, the introduction of a certain tool in the humans-with-media system impels changes in the activity, which means that the collective of humans-with-media changes according to the type of media that it incorporates. So, this notion allows overtaking the dichotomy which prevails on the subject of internal/external representations. Both constructs are so interconnected that the duality no longer makes sense and the frontier of the cognising subject is not clear at all.

It is important to note that the same authors consider that the previous experiences with certain technological tools are part of the system humans-with-media (Borba & Villarreal, 2005). By observing some students, during a certain experimental activity with a graphic calculator, they witnessed that those students used metaphors related to their background experiences with other technological devices (e.g. students mentioned that the zoom of the calculator operates the same way as the zoom of a microscope). Borba and Villarreal (1998) thus conjecture that the previous activity and the resulting knowledge with such technology (microscope) have integrated the humans-with-media unit.

This means that different collectives of humans-with-media originate different ways of thinking and knowing; for instance, the mathematical knowledge produced by humans-with-paper-and-pencil is qualitatively different from that produced by humans-with-a-spreadsheet which, in its turn, is also different from that produced by humans-with-dynamic-geometry.

This brief theoretical discussion aimed to offer an insight into the production of mathematical knowledge by means of technological tools, assuming that the

5.1.3 Humans-with-Media Mathematising

Several authors have proposed conceiving mathematics as being essentially a human activity. Hersh (1997) is convinced that all mathematics results from social phenomena and it soaks up the culture and history of humankind to the point that it is only possible to understand its essence in the light of the surrounding social contexts. Freudenthal (1973, 1983), whose ideas are the foundational basis of the Realistic Mathematics Education (RME) perspective, considered mathematics as a human construction and recognised the role of common sense as fundamental in that activity; that is to say, the experience of the individual and their interpretation of reality are two fundamental features in the construction of mathematical knowledge.

These features have a strong presence in the solutions developed by the participants in this study while solving each of the problems of the competitions. In fact, the way they unveil and build the path to the solution is often deprived of formal mathematical techniques, highlighting their vision and interpretation of the problem from the point of view of their daily common experience, which involves, as well, their use of technological tools. The competitors use their informal mathematical knowledge, expressed through what the researchers call mathematical thinking (Mason, Burton, & Stacey, 1982), which, according to Schoenfeld (1994), includes the development of a mathematical point of view and of mathematical skills, by using tools. This engagement that Schoenfeld advocates as necessary to the practice of mathematical thinking is consistent with what Freudenthal (1973) named mathematising processes—reflections on reality which lead to its understanding and alteration, through the (re)construction and the (re)organisation of mathematical contents or methods. In short, these procedures allow describing phenomena in mathematical terms, in order to deal with reality and accordingly be able to act upon it.

Based on Freudenthal, Treffers (1987) introduced two concepts which were recognised and integrated in the RME perspective: horizontal mathematisation consists of the process of exploring and interpreting real situations and problems, which lead to the establishment of mathematical concepts, while vertical mathematisation arises as a formalisation and relationship among these concepts, through its organisation, classification and generalisation. Thereby, mathematics arises as a natural process in which mathematicians interpret and organise reality according to their needs and preferences. In the same way, it is possible to state that children mathematise, that is, they reinvent mathematics in their own ways, according to their individual characteristics and under the influence of the environments in which they are immersed, currently being technologically rich and diverse.

Chapter 4 (see Sect. 4.2) considered problem-solving as an activity of mathematisation and discussed how solving contextualised problems promotes the development of conceptual models, i.e. models that have underlying mathematical ideas and relations and which facilitate the construction and development of a problem-solving strategy. According to RME, the models are "representations of problem situations, which necessarily reflect essential aspects of mathematical concepts and structures that are relevant for the problem situation, but that can have different manifestations" (Van den Heuvel-Panhuizen, 2003, p. 13). Hence, these conceptual models can comprise a textual or oral description of the problem conditions, diagrams, drawings or tables or expressions which involve mathematical symbols. These models of specific problem contexts appear, initially, as a way to represent the problem and to provide a meaning for it (Gravemeijer, 2005), and they can consist of informal strategies, based on common sense knowledge and on the student's experience.

Nevertheless, the role of these models is likely to change because as young people are becoming used to similar problems, they can start focusing on the mathematical objects, relations and procedures that characterise vertical mathematisation. The model is no longer being used solely to represent the situation; it becomes the basis of a mathematical reasoning which focuses itself on the relations involved, in a way uprooted from the context. As it is described by Gravemeijer (2005), "a model of informal mathematical activity develops into a model for more formal mathematical reasoning" (p. 95). While in the construction of an informal model, the student focuses on the relation between the contextualised situation and the mathematical procedures or concepts, the development of a formal model involves moving away from the context, focusing on the search of symbolisation, of more formal relations and strategies, which support reasoning. The conceptual model assumes, gradually and progressively, "a more object-like character" (p. 95) modifying itself until it has "a life of [its] own" (p. 98).

It is in this sense that we intend to analyse the development of conceptual models by young participants; that is, we aim to know to what extent the recognition of the affordances of a technological tool is crucial in the construction of a model during the problem-solving process, which eventually acquires "its own life" in the sense proposed by Gravemeijer.

5.1.4 Mathematisation with Dynamic Geometry Software

Dynamic geometry software (DGS) has brought a new impetus to the introduction of digital technologies in the teaching and learning of mathematics, due to its promising potentialities in geometry education (Laborde, Kynigos, Hollebrands, & Strasser, 2006; Watson, Jones, & Pratt, 2013).

Two main aspects come into play regarding the development of geometrical thinking in educational settings: the spatial aspects, which include spatial thinking and visualisation, and the aspects that relate to the ability of reasoning with theoretical concepts from the field of geometry, also including deductive reasoning

5.1 Main Theoretical Ideas

(Watson et al., 2013). Following Battista's (2007) perspective, spatial reasoning "provides not only the 'input' for formal geometric reasoning, but critical cognitive tools for formal geometrical analyses" (p. 844). By spatial reasoning, Battista means "the ability to 'see,' inspect, and reflect on spatial objects, images, relationships, and transformations" (p. 843). In addition, Watson et al. (2013) clarify that "spatial reasoning is a form of mental activity which makes possible the creation of spatial images and enables them to be manipulated in the course of solving practical and theoretical problems in mathematics" (p. 96). The authors emphasise two main activities when tackling geometry problems that often require such capability—to create and to manipulate.

Diagram generation and exploration are frequently recognised as the main affordances of dynamic geometry software that are now more powerful than ever, allowing the rapid combination and connection between geometry, measurement and algebra objects. These environments not only allow the construction of drawings, which can relate intimately to the correspondent figures due to the set of theoretical geometry rules incorporated in the DGS, as they also urge manipulation in that it becomes difficult for the user not to respond to the numerous possibilities for actions: to drag and to test, to conjecture and to verify.

In fact, dynamic geometry software can serve a double purpose: while it helps geometry concepts and ideas come to life by providing them with contextualised meaning, it may also guide students on a journey from informal ideas to more formal geometrical notions. The first purpose is well described by Leung (2008) when he states that one of the DGS main features that makes them so appealing for classroom use is to "visually make explicit the implicit dynamism of thinking about mathematical geometrical concepts" (p. 135). Accordingly, research has shown that dynamic geometry software is helpful in visualising geometric concepts and understanding rules, but also in making conjectures and generalisations, and finding relations among concepts (Baccaglini-Frank & Mariotti, 2010; Jones, 2000).

What is more, dynamic geometry software seems to play an important role when a modelling approach to a geometrical situation is required. It stimulates a movement from informal, context-dependent thinking to a more formal type of thinking, which results in the development of conceptual models. Several studies have shown that, when solving a problem using a DGS, students perceive and make use of the software affordances in order to model the situation: they undertake a construction, revealing how they are interpreting the problem and depicting some of the mathematical relationships underneath; they then explore and investigate properties, which may drive transformations at the level of their reasoning processes, hence, of their geometrical thinking (Holzl, 2001; Iranzo & Fortuny, 2011; Jones, 2000; Mousoulides, 2011). A particular study, reported by Chen and Herbst (2013), aiming to understand how the interaction with diagrams influenced students' capacity to evaluate the reasonableness of a conjecture, elected diagrams as "key resources in students' geometrical reasoning" (p. 285). They also stated that "different kinds of interaction with diagrams may engage students in particular ways of thinking" (p. 286).

Manipulating the geometrical objects on a DGS, whose rules were set upon Euclidean geometry, seems fundamental in revealing the "implicit dynamism" mentioned by Leung (2008). Dragging is one of the most studied affordances of these dynamic environments, and nowadays, it is known that it inspires conjecture generation and, therefore, may activate geometrical thinking (Baccaglini-Frank & Mariotti, 2010; Leung, 2008). This activity of conjecture generation is triggered by the observation and perception of the properties of a certain figure that remain invariant under dragging and culminates with the establishment of connections to the theoretical geometrical concepts. However, looking for patterns and for invariants, considered "a major activity in mathematical thinking" (Leung, Baccaglini-Frank, & Mariotti, 2013, p. 440) and the very "essence of dynamic geometry environments" (Laborde, 2005, p. 22), demands a combination of empirical observation and theoretical ideas that may stimulate the necessity of a proof of the emergent geometrical properties. As regarding the proving activity, one should expect the production of a "sequence of statements that logically justifies a conclusion as a consequence of the 'givens'" (Battista, 2007, p. 853). However, it has been reported that the easy and rapid production of a large amount of verifications afforded by DGS can lead students to be satisfied by "quasi-empirical arguments" (Watson et al., 2013; Holzl, 2001), especially when the conjectures to be proven arise from measurement activities.

Previous work by Jones (2000) also draws on the idea of moving from informal knowledge to formal geometrical-driven notions and relations. The author reports a study where he analysed students' given interpretations and explanations for geometrical properties of figures constructed with a DGS. The research pointed out a change in the students' patterns of thinking which were developed from "imprecise, 'everyday' expressions, through reasoning mediated by the software environment to mathematical explanations of the geometric situation" (p. 80). Since the teaching unit built to support the classroom's work had a design that favoured progressive mathematisation, the study showed that different students were able to progressively mathematise by means of a DGS.

These findings are consistent with the assumption that the malleability of DGS environments allows students to solve geometry problems, despite their different levels of mathematisation. Different humans-with-DGS are likely to produce different mathematisation when they engage in solving a geometry problem. Such differences may be related to their individual geometrical thinking skills, whereas the perceiving of the actions afforded by the software also plays an important role in that process. Bearing this in mind, we argue that dynamic geometry software allows each student, in their own level of mathematisation, to undertake particular actions, build distinctive pathways and find different results, while effectively solving the problem.

5.2 Context and Method

From the theoretical discussion included in the previous sections and from the wider framework developed in Chap. 4 (see Sect. 4.5), it is possible to think of these young participants as humans-with-media, since they convey a set of experiences and knowledge related to the use of a variety of technological tools.

Based on that assumption, we aim to understand, to a greater extent, the nature of mathematical problem-solving activity with a DGS, in a beyond-school context,

5.2 Context and Method

conjecturing that the effectiveness of each solver bears a tight relationship with their perception of the affordances of the technological tool. We analyse, at first, the recognition of the affordances of GeoGebra by several participants and then understand how this perception supports and shapes the development of suitable conceptual models which allow solving a geometrical invariance problem and communicate its solution.

By analysing the affordances of a dynamic geometry program, such as GeoGebra, in what concerns the competitors' aptitudes in the competitions, we are driven to the identification of the program's characteristics that inform the users about its potentialities of interaction, that is to say, the signals in the interface that inform the user about what can be done and how it can be done.

Having in mind the participants' productions and the general knowledge about the program, six possibilities for action are identified in GeoGebra, as a geometry exploring environment, and some examples, which help to clarify the designations used, are presented, without any intention of thoroughly listing a complete set of characteristics of the program—these are shown in Table 5.1. This categorisation serves the purpose of organising the analysis of the solutions of a geometrical problem, sent by the competitors who used GeoGebra in some point of their processes.

The empirical data used in this study come from the collecting of digital answers sent by the participants in the SUB14 (7th and 8th graders) to a geometry problem, whose context refers to the changing of the triangular shape of a certain flower bed and to the effect of this change on the area of the triangle (see Fig. 5.3).

We collected all the solutions to this problem sent by the participants as well as the e-mails exchanged with the organisation of SUB14, which contained an appreciation of the work submitted.

An initial analysis of the 227 solutions obtained allowed an organisation according to their correctness and still according to the file format which was used as a basis for the work submitted by the participants. Then, we selected the answers of the participants who used GeoGebra at some point during their problem-solving process. Afterwards, we produced a more refined analysis of these cases by describing the strategies used based on all the information available, namely, the descriptions they sent of their own reasoning and the GeoGebra files that contain the construction protocol. Notably, the construction protocol allows us to "rewind" the processes back from the final product and observe, step-by-step, the sequence of procedures used by the participants.

Table 5.1 Possibilities of action with GeoGebra

	Affordance	Examples
IC	Immediate constructions	Place a point/draw a segment
M	Measurement	Length/perimeter/area
RC	Referential constructions	Draw a straight line perpendicular to a segment/ draw a circumference with a radius of 2 cm
SP	Setting properties	Bold, dotted, change the colour, hide an object, define rounding of measure units, use labels
PC	Constructions using parameters or variables	Use a selector to make the length of a segment vary
DE	Drag and explore	Drag a point and exploit the geometrical properties

Rose explained to her gardener that she would like to make a triangular area of flowers in her rectangular lawn garden. The worker picked up a stick 2 meters long, stretched it perpendicular to one of the sides of the garden, in a random point. Then, with a string, drew a line through the end of the stick (F) and joining the two opposite sides of the rectangle, getting, this way, the inner triangle [EGH].

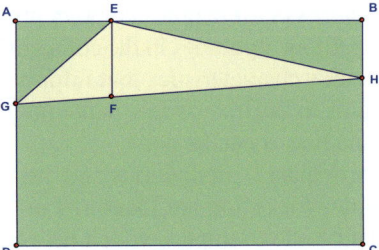

The following day Rose looked at the triangle and did not like it, she displaced the stick to another point at random on the garden edge and she got a different triangle [EGH].

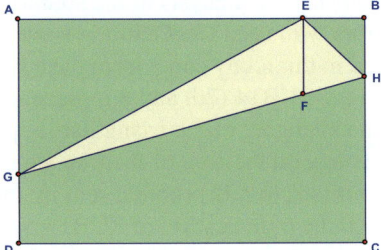

When the gardener came back he protested, saying that the area for flowers had diminished. But Rose assured him that it did not. Who is right and why?

Do not forget to explain your problem solving process!

Fig. 5.3 Problem #6 from the SUB14 competition (edition 2010–2011)

5.3 Data Analysis

In this section, we present the analysis of the solutions of a geometrical problem proposed in 2010–2011 in the SUB14, which directs the exploration of the invariance of an area, by changing the geometrical figure.

5.3.1 The Problem: Building a Flowerbed

When this problem was proposed, there were 126 students from 7th grade and 101 students from 8th grade persisting in the competition. From these, some 67 competitors from 7th grade ($\approx$53 %) and 35 from 8th grade ($\approx$35 %) answered correctly. The number of participants that did not try to answer the problem or sent incorrect answers (about 55 % of the total), together with the feedback that was sent asking for an explanation for the strategy used or a justification of the solution, is a strong sign of the difficulties felt by the competitors in this problem.

If we consider only the solutions accepted as correct, we observe that the competitors used a range of tools to solve and express the given problem. However, we notice that the use of a text editor is predominant (used by 47 % of the competitors) to paste pictures from the statement or to take advantage of the automatic shapes, text boxes, brackets and mathematical symbols or just to explain their procedures in writing. Almost one-third of the participants preferred to use only an electronic message, so they typed a description of their reasoning and presented the calculations required directly on the message box. The percentage of competitors from 8th grade who preferred to send the answer in that way is roughly double the percentage found in 7th grade.

We must also highlight that around 14 % of the competitors chose to send a picture showing the solution, resulting from a digitalisation of their paper-and-pencil work or showing a geometrical construction with an explanation of their reasoning created on a picture editor (e.g. paint). This percentage is higher in the 7th graders. There were also two competitors who presented their solution through a presentations editor and one in PDF (though each seems to have used a text editor to compose it). A total of nine 7th graders, some organised in small teams, submitted a GeoGebra file containing their solution to this problem. Since two files convey similar approaches, we selected the remaining four solutions for a deeper analysis which we present in the following section.

5.3.2 Zooming in: The Participants' Productions

In this section, we present the four solutions to the problem "building a flowerbed" produced by seven participants (two teams, one with two and the other with three students, and two competitors who applied individually), all of them attending the 7th grade. These participants used GeoGebra in at least one of the different steps of their approach to the problem; hence we aim to describe in detail the main features that characterise their activity.

5.3.2.1 Exhibit A: The Solution of Marta and Miguel

Marta and Miguel sent an initial answer to the problem, presenting their reasoning in a text document. As it was not correct, the organising committee sent them back an appreciation concerning the triangle analysis obtained when a triangular flowerbed is divided in two triangles, using the stick. The second answer received consisted of a file in GeoGebra format (Fig. 5.4).

The construction protocol allowed us to reconstruct the participants' procedures during the initial task of reproducing the statement conditions (Table 5.2).

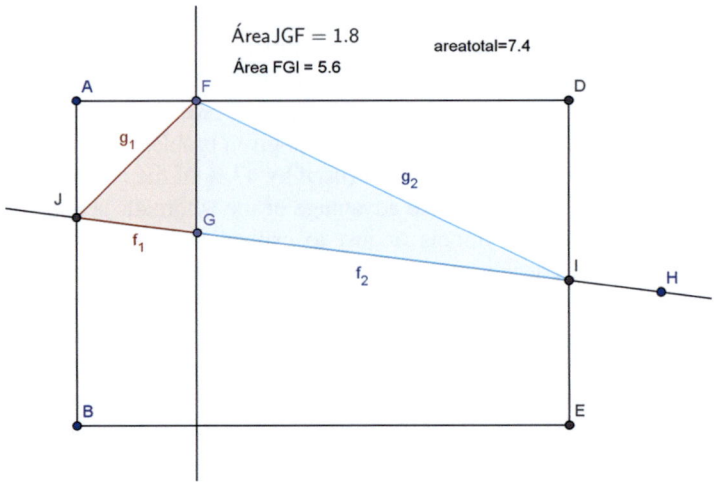

Fig. 5.4 Construction in GeoGebra, sent by Marta and Miguel

Table 5.2 Sequence of actions and affordances perceived by Marta and Miguel

Actions	Affordances
Place A, B and draw segment AB	IC
Place C, exterior to AB	IC
Draw b, a line parallel to AB passing through C	RC
Draw c, a line perpendicular to b passing through A	RC
Draw d, a line parallel to c passing through B	RC
Find the intersection of b and c; b and d, and name it D and E, respectively	IC
Build segments DE, AD and BE	IC
Hide the lines b, c and d	SP
Place F on AD	IC
Draw h, perpendicular to AD passing through F	RC
Draw circumference of centre F and radius 2 units	RC, M
Find G, one of the intersections of the circumference and h	IC
Place H on the right exterior side of the drawing, build j passing through G and H	IC, RC
Find the intersections of DE and j, and name it I, and of AB and j, and name it J	IC
Build the polygons JGF and FGI, changing colours	IC, SP
Measure the area of each polygon	M
Determine the sum of the two areas and name it "areatotal"	M, SP

This process shows that the competitors recognise and respond to GeoGebra's invitations to action, namely, to represent rigorously the conditions of the problem and to obtain a solution by calculating the areas of the considered regions. Marta and Miguel conclude that Rosa was right and that the area would not change, so this finding may have been achieved by dragging the vertex F, simulating the changing in the stick's position and checking that the total area is invariant despite

5.3 Data Analysis

the alterations in the partial areas. They fail to submit a mathematical justification for the invariance of the area, which may result from the "certainty" they seem to get from dragging F.

We observe that these youngsters use GeoGebra with the purpose of elaborating a representation of the situation that allows them to manipulate and measure; i.e. it supports the construction of a conceptual model of a particular situation. The effective recognition of some of GeoGebra's affordances impels an activity based on representation, observation and interpretation of the situation, so that it enables dealing with the problem, in a reasonably effective manner. This approach discloses a horizontal mathematisation activity, in which more elementary concepts are put in action through objects that acquire dynamism, converging in the organisation and development of a conceptual model. The solution presented by this team is based on the construction of an "informal model", where they use mathematical concepts and procedures that "become alive" through the construction and manipulate them to infer the solution. As these competitors did not present any attempt to justify their conclusions, we consider that GeoGebra was used to bring to life the conceptual model that supports the problem-solving, with the purpose of obtaining the solution.

5.3.2.2 Exhibit B: The Solution of Andreia, Lucas and José

Andreia, Lucas and José sent a solution that includes a GeoGebra construction, also allowing an analysis of the corresponding construction protocol. They included a short text in order to justify the conclusion which was obtained through the manipulation of the construction (Fig. 5.5).

The construction process carried out by this team differs slightly from that described previously, mainly in terms of the sequence of actions (Table 5.3).

The short text (Fig. 5.6) sent by the competitors reveals their attempt to justify the invariability of the area that they were observing: "triangles with the same base and the same height have equal areas". That conclusion arises from the manipulation of the vertices E and G "to match situations described in the problem".

This solution provides another example of how the recognition of GeoGebra's affordances drives a rigorous representation of the context described, to reach the solution. These competitors are able to perceive a wide range of affordances and to

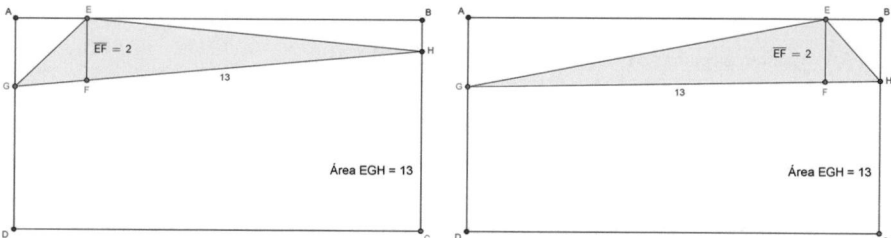

Fig. 5.5 Two instances of a dynamic geometry construction sent by the team of Andreia, Lucas and José

Table 5.3 Sequence of actions and affordances perceived by Andreia, Lucas and José

Actions	Affordances
Place A, B and draw segment AB	IC
Place E on AB	IC
Draw circumference of centre E and radius 2 units	RC, M
Draw h, perpendicular to AB, passing through E	RC
Draw b, perpendicular to AB, passing through B	RC
Place C on b	IC
Draw c perpendicular to AB passing through A	RC
Draw d, parallel to AB passing through C	RC
Find D as the intersection of the lines c and d	IC
Build segments BC and DC	IC
Find I and F as the intersections of h and the circumference	IC
Build the segments AD and EF (the stick)	IC
Measure the distance between E and F	M
Place G on AD and draw j, a line passing through G and F	IC
Name H to the intersection of j and BC	IC
Build the segments GH, EG and EH	IC
Build the polygon EGH, obtaining its area	IC
Measure the area of the polygon EGH	M
Hide lines and points that are no longer needed	SP

Resposta:

Triangulos com a mesma base e a mesma altura têm áreas iguais.
No seguinte ficheiro geogebra movendo unicamente os pontos E e G para as situações descritas no problema, facilmente se constata que o valor da área se mantém inalterável. Por isso o jardineiro não tem razão quando diz que a área diminuiu, pois a área mantém-se igual.

ANEXO (ABRIR NO PROGRAMA GEOGEBRA)

Answer:

Triangles with the same base and the same height have equal areas.
In the following geogebra file moving only the points E and G to match the situations described in the problem, it is easily verified that the area remains unchanged. That is why the gardener isn't right when he says that the area is smaller, because the area remains the same.

ATTACHMENT (OPEN WITH GEOGEBRA)

Fig. 5.6 Written explanation sent by Andreia, Lucas and José

take advantage of them to construct their own conceptual model, by resorting to one step constructions and referential constructions, by setting properties of geometrical objects, by using measurement tools and by dragging objects.

A closer and attentive look offers some evidence about how the activity with the technological tool can influence the development of a "model of this specific situation", mainly because these youngsters are already trying to include a mathematical explanation in their resolution. Solving this problem, that is to say, developing this

5.3 Data Analysis 129

conceptual model, is not confined to the representation of the statement conditions, but, beyond that, it brings together the geometric representation, the manipulation and the search for a justification, whose necessity seems to be stemming from the verification of the invariance of the area.

The "invisibility" of the mathematical ideas, mentioned in the literature, is noticeable in this production. The competitors naively accepted the result given by GeoGebra and used it to attempt a mathematical justification without a deliberate analysis of such an outcome. In fact, dragging E and G causes the segment GH to change its length, but the measure given by GeoGebra remains the same. This is, probably, due to the type of rounding established—to the unit in this case—and, therefore, it is likely that Marta and Miguel did not realise it. They lack a certain critical sense in their analysis of the digital representations affecting their capacity to transform information into knowledge (Noss, 2001). Overall, it is the use of the technological tool that supports the search for a mathematical sense for the problem and for the answer obtained, so we can consider that these participants are using GeoGebra to interpret the solution.

5.3.2.3 Exhibit C: The Solution of Sara

Sara appears to have experienced some difficulties solving this problem, especially when she needed to "put it in words". That is the reason why she decided to complete the justification by sending a picture (print screen) of her construction in GeoGebra. As she did not send the GeoGebra file, it was not possible to use the construction protocol to reconstruct her procedures. Nevertheless, we included the analysis of this case because the participant's description of her reasoning is quite detailed. In any case, it is not possible to present a list of her sequence of actions and the corresponding affordances, as in the previous examples.

As she explains (see Fig. 5.7), Sara starts by "imagining" that the rectangle would be 12 cm length (which she names width) and she creates a representation of a lawn garden and a flowerbed, following the statement conditions.

Based on the construction on the left of Fig. 5.8 and some calculations to determine the area of the triangle, Sara observes that this value matches with that she attributed to the length of the side of the rectangle (Fig. 5.9). She aligns the larger side of the triangle EFG with the larger side of the rectangle, so that they are parallel, suggesting that it is a reference position, allowing her to test other positions.

When Sara proceeds to the second construction (Fig. 5.10), besides changing the position of the triangle, she divides the triangular flowerbed into two interior triangles, ONM and OMK, indicating that she is trying to make sense of the invariance of the area and is looking for a plausible mathematical reason. Although she claims to have determined those two areas, the picture sent neither contains its explicit calculation using the measurement tools nor its result.

Sara explains that the 2 m long stick represents the base of each of these triangles, ONM and OMK, and she reproduces their heights (relative to the vertical stick) through two segments, which are identified as a1 and b1. She states that the

Resposta:	Answer:
A senhora Rosa é quem tem razão, a area dos triangulos é a mesma. Para explicar também enviei em anexo. Fui imaginar que o rectangulo tinha 12 cm de largura, a formula da area de um triangulo é base x altura a dividir por dois. No primeiro triangulo a area é 12, porque a sua base é 12, 12x2=24 e 24 a dividir por 2 vai dar 12. Area=12 No segundo dividi-o em dois triangulos, onde calculei a area de cada um e depois somei e voltaria a dar 12, desenhei uma linha recta correspondente a area de cada parte do triangulo que dividi em dois, depois juntando as duas rectas, com uma regua medi e dava 12, concluindo assim que era verdade. Também usei o geogebra para resolver, como era difícil explicar por palavras reduzi a este pequeno texto e enviei um anexo com mais das explicações.	Mrs. Rosa is right; the area of the triangles is the same. I sent an attachment that explains. I imagined a rectangle with 12cm of width, the formula of the area of a triangle is base x height divided by two. In the first triangle the area is 12, since its base is 12, 12x2=24 and 24 divided by 2 equals 12. Area=12 In the second one, I divided it into two triangles, I calculated the area of each one and then added it up and it totals 12; I drew a straight line with a length corresponding to each of the areas of the two parts of the triangle that was divided into two, then by joining the two lines, I measured it, like with a ruler, and I saw it measured 12, thus concluding it was true. I used geogebra to solve it and since it was difficult to put it in words, I just summarised this small text and sent the attachment with more explanations.

Fig. 5.7 Written explanation sent by Sara, by e-mail

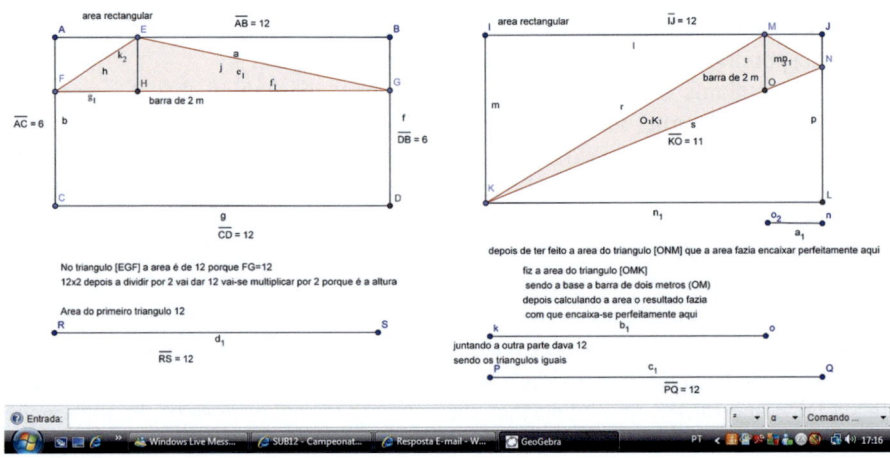

Fig. 5.8 Picture sent by Sara

5.3 Data Analysis

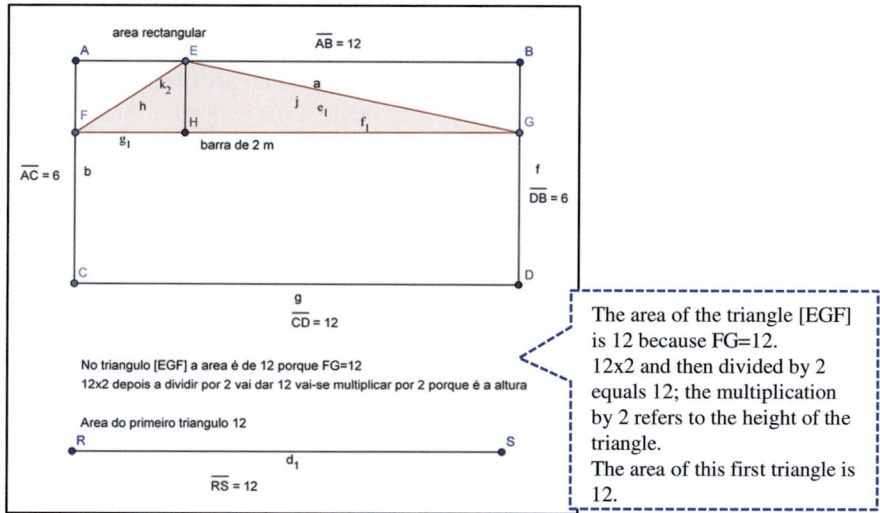

Fig. 5.9 Extract of the picture sent by Sara, in detail

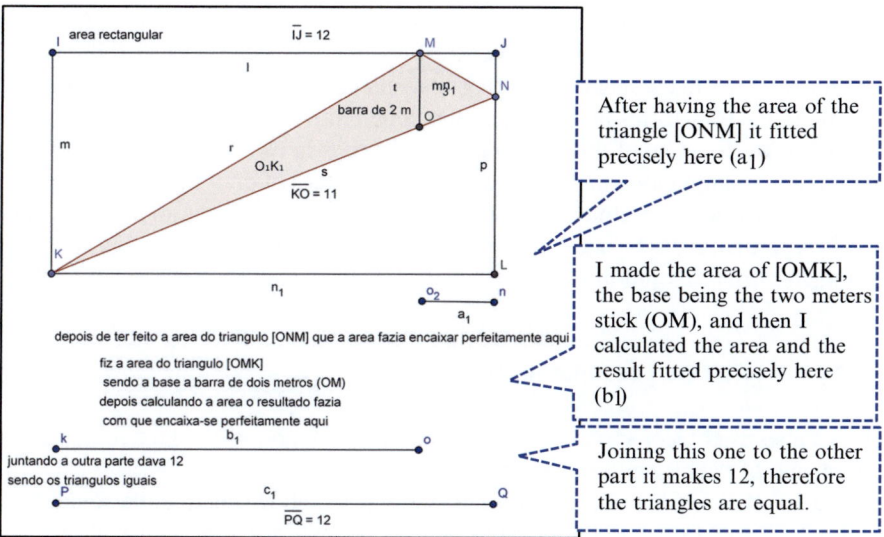

Fig. 5.10 Extract of the picture sent by Sara, in detail

area of each of the triangles has precisely the same value as the length of each of the segments and justifies it with the length of the altitude being 2. Finally, she refers that by "joining" the two segments—that is to say, the heights of the two interior triangles—the result is 12 cm, the same as the length of the rectangle. As the values of the lengths are equal to the values of the areas, then the area of the triangular flowerbed will always be 12 (the length of the rectangle), independently of the position and the shape of the flowerbed.

This case shows a participant who, similarly to the previous cases, recognises a significant amount of affordances of GeoGebra and is able to answer to those invitations to action in order to represent a particular case of the situation described in the problem. As the young girl uses the tool to build a model with the concrete dimensions that she chose, the solution found is also concrete and specific, as it corresponds only to that particular problem.

The development of this conceptual model is promoted, as happened in the previous cases, by the recognition of the affordances of GeoGebra. Nevertheless, in the second part of her work, Sara continues to be engaged in the use of GeoGebra to draw a convincing justification, and this activity shows her own way of thinking about the problem with this tool and, particularly, about the conclusion attained in the first approach. In a way, this case shows a solving-and-expressing activity as the participant understands the need mathematically to validate her observation and undertakes the development of the conceptual model instantiated from elementary mathematisation processes.

This conceptual tool, developed in close articulation with the technological tool, provides "explicit descriptive or explanatory systems" (Lesh & Doerr, 2003, p. 18) which reveal the mathematical thinking that shapes the interpretation of the result (solution) and transforms this production in an effective model.

As the development of this conceptual model runs in two moments—one of representation, manipulation and verification and the other of drawing a mathematical justification, which can almost be considered a geometrical proof, though it is centred in a particular case—we consider that this use of GeoGebra contains an intention to confirm the solution.

5.3.2.4 Exhibit D: The Solution of Jessica

Jessica also used GeoGebra to reproduce the construction of the rectangular lawn garden and of the triangular flowerbed (Fig. 5.11).

The text that Jessica sent in her e-mail (Fig. 5.12) allows us to understand thoroughly how she arrives to the recognition that the area of the triangular flowerbed matches the value chosen to the length of the side of the rectangle. The movement of the "stick" along the side of the rectangle was decisive in understanding that it divides the triangular flowerbed in two triangles.

The GeoGebra construction shows that Jessica was working from the point of view of the geometric properties and of the relationships that the conditions of the statement impose, rather than with the purpose of measuring or calculating. This is

5.3 Data Analysis

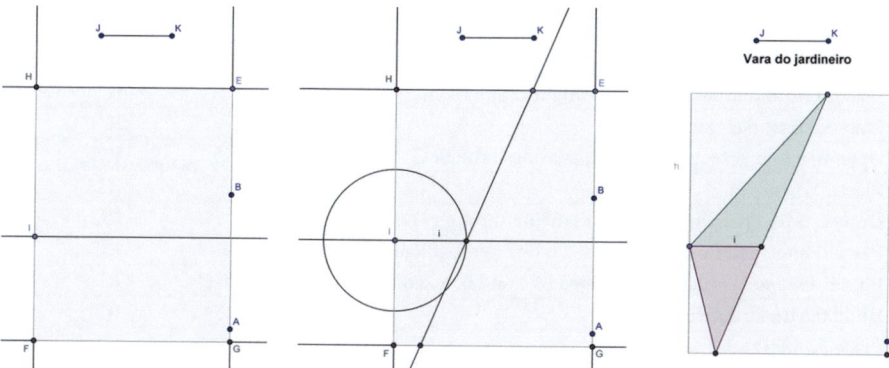

Fig. 5.11 Three steps of Jessica's construction

Resposta:
O triângulo amarelo (zona de flores) está dividido em dois triângulos pela vara de 2 metros que o jardineiro colocou. Sabemos que a base desses dois triângulos mede 2 metros - o comprimento da vara.
Para medir a área de um triângulo, fazemos a seguinte conta: altura x base / 2
Para medir a área desses dois triângulos, será então altura x 2 /2. Ora, está claro que 2 / 2 = 1, portanto, a área desses dois triângulos é igual à sua altura.
Podemos afirmar que a soma das alturas dos dois triângulos é igual ao comprimento do rectângulo (jardim de relva). Portanto, a área da zona das flores é igual ao comprimento do jardim de relva rectangular.
Se o comprimento do rectângulo (jardim de relva) não muda, então a área do triângulo (zona de flores) também se mantém. Por outras palavras, a Rosa tem razão.

Answer:
The yellow triangle (flowerbed area) is divided into two triangles by the 2 m stick placed by the gardener. We know that the base of those two triangles measures 2 meters - the length of the stick.
In order to measure the area of a triangle, we use the calculation: height x base / 2
In order to measure the area of those two triangles we will have height x 2 / 2. Well, it is clear that 2 / 2 =1, therefore the areas of those two triangles are equal to their heights.
We can say that the sum of the heights of the two triangles is equal to the length of the rectangle (lawn garden). Therefore the area of the flowerbed is equal to the length of the rectangular lawn garden.
If the length of the rectangle (lawn garden) doesn't change, then the area of the triangle (flowerbed) remains unchanged. In other words, Rosa is right.

Fig. 5.12 Written explanation sent by Jessica, by e-mail

the work of a participant who recognises a large range of GeoGebra affordances; namely, she shows she is able to respond to the invitations to action in order to establish her solution (Table 5.4). This case sustains, once more, the strong belief that problem-solving is an activity of permanent solving and expressing, supported by a multiplicity of "descriptive or explanatory systems" (Lesh & Doerr, 2003, p. 18).

According to the way she expresses herself in written words, it is equally recognised that the construction and the manipulation of the construction, enhanced by the use of GeoGebra, motivates the development of a productive way of thinking

Table 5.4 Sequence of actions and affordances perceived by Jessica

Actions	Affordances
Place A, B and draw a line a passing through them	IC
Place C on a, but exterior to AB	IC
Draw b, a line perpendicular to a passing through C	RC
Place D on a	IC
Draw c, a line perpendicular to a passing through D	RC
Place E on a, and through it draw d a line perpendicular to a	RC
Find F, G, and H—the intersections of c and b, a and b; c and d, respectively	IC
Build the quadrilateral CEHD	IC
Place I on HD	IC
Draw f a perpendicular line to HD passing through I	RC
Place J and K outside the rectangle and build the segment JK	IC
Draw a circumference with centre I and radius equal to the segment JK (slider)	PC
Find L, one of the intersections of the circumference and f	IC
Build the segment LI, the stick	IC
Place M on HE and draw the line j passing through L and M	IC
Find N, the intersection of j and DC	IC
Build the polygons NIM, ILM and ILN, colouring their interior	IC, SP

about the problem. Consequently, the organisation and the interpretation of the situation, from a mathematical point of view, that is to say, the horizontal mathematisation of the situation, triggers the development of a very powerful conceptual model that is quite different from those previously analysed both because they have a less sophisticated structure or they are quite informal and because they are situation framed. The fact that Jessica recognises a particular affordance of this dynamic geometry environment—using parameters or variables in the constructions—grants the simulation of a slider that controls the length of the stick. This new geometric element spurs a new activity, which is the analysis of a variable that is not explicit in the statement of the problem; hence the participant's exploration goes far beyond what was requested in the problem.

During Jessica's construction activity, GeoGebra acquires the role of a tool-to-think-with rather than that of a tool-to-calculate-with as there are no evidences of Jessica having used the measurement tools. The solution is complete after the inclusion of a detailed justification, where she explains her reasoning and points out the calculations she considers necessary, those that aim "to demonstrate" the truthfulness of her general finding. Although the construction respects the statement conditions, the use of a segment that simulates a slider, the existence of a moveable point located near the right inferior vertex of the rectangle which allows the regulation of the dimensions of the rectangle and the absence of measures and calculations are signalling an intention to formalise and generalise. The manipulation of the variable "height" of each smaller triangle that results from changing the position of the stick suggests a vertical mathematisation activity in which the student develops a textual proof. She apparently translates the change of the heights to an algebraic recognition of their role in the area of the flowerbed.

If, at a first stage, the invariance of the area may not be obvious to the students, the representation of the situation through GeoGebra triggers the emergence of a conceptual model that comprises, simultaneously, the perception of the stick as the shared base of the two smaller triangles and the calculation of the corresponding areas. It seems that the interactive activity with the tool, that is to say, the manipulation and observation, supports the development of a conceptual model to solve the problem, with outlines of a formal mathematical model. So, in this case the construction performed with GeoGebra induced obtaining a proof of the solution.

5.3.3 Zooming Out: Comparing and Contrasting

Looking back at these solutions from a wider perspective allows the identification of common general features, as well as of the singularities in these young competitors' activity, while they appropriate the problem and try to develop a productive way of thinking about it.

These four cases, analysed in the light of the metaphor "humans-with-media" (Borba & Villarreal, 2005), illustrate some symbiotic relationships among the youngsters who solve the problems and the technologies in use. Their activity results from a combination between mathematical knowledge and procedures, particularly geometrical and analytical, and knowledge about the tools used, especially about GeoGebra but also the e-mail and the text editor.

The conscious option for this dynamic geometric environment, to solve this particular problem, is not detached from the purposes and intentions of each student. In fact, the kind of activity that GeoGebra is enabling does not replace, nor simply complement, any other kind of activity, for instance, a paper-and-pencil solution. It results rather from the expectation these competitors have that GeoGebra offers them the possibility to give life to the static model presented in the problem, supporting the development of a solving strategy which supports them, with some certainty, to achieve the solution.

The high proficiency of these students-with-media is revealed in the way they answer the emergent invitations to take action with GeoGebra. They perceived, at first, the convenience and relevance of the dynamic geometry environment to represent the situation described in the problem, foreseeing that the reproduction of the frozen images and ideas presented in the statement would transform them into dynamic images and ideas (Leung, 2008), providing a greater freedom in their manipulation.

These youngsters' familiarity with the software is quite visible as all of them demonstrate the capability to recognise a diversity of GeoGebra's affordances. They all started by setting a plan which consisted of representing a rectangular lawn garden, and so they identified the GeoGebra affordances that grant them, on one hand, the perpendicularity of the sides and, on the other hand, that the opposite sides will have the same length. Besides undertaking one step constructions (points, segments and straight lines), the participants recognize the need to use referential constructions (such as perpendicular or parallel lines) to obtain a 'robust' rectangle (that means that even after manipulating the objects, the geometric features remain intentionally incorporated). In order to represent the triangular flowerbed, they use a range of quite similar affordances and define the length of the stick by using a circumference whose centre is a free point on one of the rectangle sides and whose radius is fixed, in the three first cases, and variable upon the length of a selector, in the last case. This referential construction comprises knowledge about GeoGebra's specific mechanism which allows defining distances with a concrete or variable value.

Hence, these students-with-media have a clear notion that the constructions that are convenient and useful to solve the problem are those that allow them to change the shape of the triangular flowerbed, although all the other attributes of the construction remain the same, as the garden does not change and the stick has a fixed dimension.

In their productions, another common feature is related to the solution's appearance: not all the objects which were constructed are effectively visible in the file they sent as a solution. At a certain point of the solution process, these students-with-media used the option "setting properties" to hide the geometric elements that were used as support to others, for example, the straight lines on which segments are constructed or the circumferences that are only used to set the length of the segment that represents the stick. The participants know that those auxiliary constructions are essential and they cannot be eliminated even though they have already served their function. They have a rather deep insight into GeoGebra since they perceive a double purpose in this affordance: on one side, to hide an object, which is used to generate others, allows them to clean the construction, diminishing the number of visible entities and to focus the attention on those which are really needed during the development of the strategy; on the other side, cleaning the construction makes it more elegant, simpler and more attainable to those who are going to analyse it. The "setting properties" tool, or in this concise case the "show/hide object" tool, is a resource used for solving, but it is also a very important resource for expressing the solution.

In general, all these participants combined one step constructions (points, segments, straight lines), with referential constructions (perpendicular or parallel lines, a circumference defined by a point and a radius) setting properties (to hide objects), with the manipulation of certain free objects to analyse their effect through "dragging and exploring".

Considering that, at this stage, all the participants have constructed a representation similar to the scheme presented in the statement, although dynamic and manipulable, the way they interact with it influences the development of a conceptual model. In the two first cases, the solution arises from the usage of measurement

5.3 Data Analysis

tools; consequently, measuring becomes a vital action and encompasses all the justification presented. In the third case, the conjecture, based on a particular case, drives the search for a geometrical proof, which is sought and accomplished in GeoGebra and with GeoGebra. In the last production, the incorporation of a larger degree of freedom in some objects triggers a more global understanding of the problem and leads to the search of a general solution, a proof.

The absence of a justification or a proof of the solution obtained may find its source in a certain "invisibility" of the mathematical ideas brought by the vast quantity of particular cases experienced and the intermediate results that GeoGebra provides. Indeed, in the first case, the invariability of the values obtained by measuring seems to have produced a sense of total evidence, thus overshadowing the need to find a mathematical proof. In contrast, the participants in the second presented case yield themselves to the evidences they observe from the manipulation of the construction and seem to accept naively that the length of the longer side of the triangular flowerbed does not change. This idea emerges from their written observation regarding the areas of triangles with the same base and height and the fact that, in the GeoGebra construction, the length of the longer side of the triangular flowerbed maintains under dragging. So they only use this result in their assessment, without trying to justify it in a mathematical way, apparently lacking the sense of critical analysis of the digital representations, as referred by Noss (2001), which influences their capacity to transform the information obtained with GeoGebra into understanding and knowing mathematically the situation.

The problem-solving activity in which all these students-with-media involve themselves present very similar features to the point that it is possible to consider that they are inseparable from GeoGebra. Solving-with-GeoGebra encloses a range of procedures which come out through the symbiosis between the affordances of GeoGebra and the youngster's aptitudes: the construction is essential to interpret the initial conditions; the dragging influences the identification of a strategy and the formulation of a conjecture; in one of the cases, the construction becomes crucial again to find out a geometrical explanation, while, in the other case, the symbolic manipulation demonstrates the conjecture.

Considering the conceptual models developed by these participants, it is possible to say that they are different, depending on the considered unit of analysis, i.e. the different collectives of "students-with-media". Nevertheless, those different conceptual models interweave some mathematical contents and procedures with some affordances of the tool. The constructions, the strategies, the findings and justifications presented in each case sum up a conceptual model of the invariance of the triangle's area and expose the activity of these students-with-media, who we could describe as "students-with-GeoGebra". Therefore, we have evidence supporting that this tool—which is used in some cases to construct and measure, in others to conjecture, to verify or to think with—influences the type of mathematical thinking produced (Villarreal & Borba, 2010) which, in turn, influences the conceptual model that is developed.

5.4 Discussion and Conclusion

The data presented illustrate the diversity of ways of thinking and modes of action: four groups of solvers, who certainly have very different learning experiences, attend different schools and live in different places, realise and recognise the potential relevance of a single tool, GeoGebra, in solving this problem.

The solutions previously described and discussed highlight different mediation features, made possible through the mathematical and technological representations, namely, in the construction of conceptual models to (a) get a solution, (b) interpret a solution, (c) check a solution and (d) prove the solution.

The data suggest that the differences found are strongly related to the dynamic nature of the mathematical representations afforded by the tool, in depicting the problem conditions. For example, the introduction of additional elements to the figure led to powerful understandings of the problem, the generalisation and formalisation. In one production, the invariance of the area is not only numerically recognised but also geometrically explained; in another situation, the free elements allow seeing the answer to the given problem as a particular case of a more general statement; yet another case draws on the geometrical configuration in order to produce a formal mathematical model of the invariance of the area.

Regardless of their prior level of mathematisation, the use of GeoGebra empowers each participant during the problem-solving activity. These students-with-GeoGebra are able to recognise a sufficiently large set of affordances that allow them to obtain an acceptable solution to the problem. Yet again, the main difference among the solutions presented lies in the relationship between the aptitude of the solvers and their perception of the affordances. Even though they all recognise the affordances of the tool, some engage in a more elementary and less elaborate activity, while others engage in a more advanced and sophisticated activity. This range is also noticeable when looking at the mathematical production itself. In particular, the conceptual models developed by these students-with-GeoGebra also range from a horizontal mathematisation, very much attached to the context and to the obvious confirmation provided by the tool, to a form of vertical mathematisation where the student abandons the situational context and proceeds to create a mathematical proof that validates the model, as shown in the last exhibit.

According to Villarreal and Borba (2010), different people with different media produce qualitatively different mathematical knowledge. We are here offering evidence that different students solving the same problem with the same media and recognising a relatively similar set of affordances of the tool produce different digital solutions, but they also generate qualitatively different conceptual models, in this case, for the invariance of the area. This distinction in terms of the mathematical thinking developed is somewhat built on the symbiotic relationship between the aptitudes of the solvers and their perception of the affordances of the tool.

The example provide solid evidence of how the spontaneous use of technology changes and reshapes mathematical problem-solving. The spectrum of the problem solutions also highlight the effectiveness of the use of digital tools to structure, support and extend mathematical thinking, meaning-making and knowledge in students' problem-solving.

References

Artigue, M. (2007). Digital technologies: A window on theoretical issues in mathematics education. In D. Pitta-Pantazi & G. Philippou (Eds.), *Proceedings of CERME 5* (pp. 68–82). Larnaca, Cyprus: Cyprus University Editions.

Baccaglini-Frank, A., & Mariotti, M. A. (2010). Generating conjectures in dynamic geometry: The maintaining dragging model. *International Journal of Computers for Mathematical Learning, 15*, 225–253.

Battista, M. (2007). The development of geometric and spatial thinking. In F. Lester Jr. (Ed.), *Second handbook of research on mathematics teaching and learning*. Charlotte, NC: Information Age Publishing.

Borba, M., & Villarreal, M. (1998). Graphing calculators and the reorganization of thinking: The transition from functions to derivative. *Proceedings of the 22nd Psychology of Mathematics Education Conference* (Vol. 2, pp. 136–143). Stellenbosch, South Africa: PME.

Borba, M., & Villarreal, M. (2005). *Humans-with-media and the reorganization of mathematical thinking*. New York, NY: Springer.

Brown, J., Stillman, G., & Herbert, S. (2004). Can the notion of affordances be of use in the design of a technology enriched mathematics curriculum? In I. Putt, R. Faragher, & M. McLean (Eds.), *Mathematics Education for the third millennium: Towards 2010, Proceedings of the 28th Annual Conference of the Mathematics Education Research Group of Australasia* (Vol. 1, pp. 119–126). Sydney, Australia: MERGA.

Chemero, A. (2001). What we perceive when we perceive affordances. *Ecological Psychology, 13*, 111–116.

Chemero, A. (2003). An outline of a theory of affordances. *Ecological Psychology, 15*(2), 181–195.

Chen, C., & Herbst, P. (2013). The interplay among gestures, discourse, and diagrams in students' geometrical reasoning. *Educational Studies in Mathematics, 83*(2), 285–307.

Drijvers, P., Godino, J. D., Font, V., & Trouche, L. (2013). One episode, two lenses; A reflective analysis of student learning with computer algebra from instrumental and onto-semiotic perspectives. *Educational Studies in Mathematics, 82*(1), 23–49. doi:10.1007/s10649-012-9416-8.

Freudenthal, H. (1973). *Mathematics as an educational task*. Dordrecht, The Netherlands: Reidel.

Freudenthal, H. (1983). *Didactical phenomenology of mathematical structures*. Dordrecht, The Netherlands: Reidel.

Gibson, J. (1977). The theory of affordances. In R. Shaw & J. Bransford (Eds.), *Perceiving, acting, and knowing: Toward an ecological psychology* (pp. 67–82). Hillsdale, NJ: Erlbaum.

Gibson, J. (1986). *The ecological approach to visual perception*. Hillsdale, NJ: Erlbaum.

Gravemeijer, K. (2005). What makes mathematics so difficult, and what can we do about it? In L. Santos, A. P. Canavarro, & J. Brocardo (Eds.), *Educação matemática: Caminhos e encruzilhadas* (pp. 83–101). Lisbon, Portugal: APM.

Greeno, J. (1994). Gibson's affordances. *Psychological Review, 101*(2), 336–342.

Hersh, R. (1997). *What is mathematics really?* Oxford, UK: Oxford University Press.

Holzl, R. (2001). Using dynamic geometry software to add contrast to geometric situations: A case study. *International Journal of Computers for Mathematical Learning, 6*, 63–86.

Iranzo, N., & Fortuny, J. (2011). Influence of GeoGebra on problem solving strategies. In L. Bu & R. Schoen (Eds.), *Model-centered learning: Pathways to mathematical understanding using GeoGebra* (pp. 91–104). Rotterdam, The Netherlands: Sense Publishers.

Jones, K. (2000). Providing a foundation for deductive reasoning: Students' interpretations when using dynamic geometry software and their evolving mathematical explanations. *Educational Studies in Mathematics, 44*, 55–85.

Laborde, C. (2005). Robust and soft constructions: two sides of the use of dynamic geometry environments. In S. Chu, H. Lew, & W. Yang (Eds.), *Proceedings of the 10th Asian Technology Conference in Mathematics* (pp. 22–36). Cheong-Ju, South Korea: Korea National University of Education.

Laborde, C., Kynigos, C., Hollebrands, K., & Strasser, R. (2006). Teaching and learning geometry with technology. In A. Gutiérrez & P. Boero (Eds.), *Handbook of research on the psychology of mathematics education* (pp. 275–304). Rotterdam, The Netherlands: Sense Publishers.

Lesh, R., & Doerr, H. (2003). Foundations of a models and modeling perspective on mathematics teaching, learning, and problem solving. In R. Lesh & H. Doerr (Eds.), *Beyond constructivism: Models and modeling perspectives on mathematics problem solving, learning, and teaching* (pp. 3–33). Mahwah, NJ: Erlbaum.

Leung, A. (2008). Dragging in a dynamic geometry environment through the lens of variation. *International Journal of Computers for Mathematical Learning, 13*, 135–157.

Leung, A., Baccaglini-Frank, A., & Mariotti, M. A. (2013). Discernment of invariants in dynamic geometry environments. *Educational Studies in Mathematics, 84*, 439–460.

Lévy, P. (1990). *As Tecnologias da Inteligência. O Futuro do Pensamento na Era da Informática*. Lisbon, Portugal: Instituto Piaget.

Martin, A., & Grudziecki, J. (2006). DigEuLit: Concepts and tools for digital literacy development. *Innovation in Teaching and Learning in Information and Computer Sciences, 5*(4), 249–267.

Martinovic, D., Freiman, V., & Karadag, Z. (2013). Visual mathematics and cyberlearning in view of affordance and activity theories. In D. Martinovic, V. Freiman, & Z. Karadag (Eds.), *Visual mathematics and cyberlearning* (pp. 209–238). New York, NY: Springer.

Mason, J., Burton, L., & Stacey, K. (1982). *Thinking mathematically*. Bristol, UK: Addison-Wesley.

Mousoulides, N. (2011). GeoGebra as a conceptual tool for modeling real world problems. In L. Bu & R. Schoen (Eds.), *Model-centered learning: Pathways to mathematical understanding using GeoGebra* (pp. 105–118). Rotterdam, The Netherlands: Sense Publishers.

Noss, R. (2001). For a learnable mathematics in the digital culture. *Educational Studies in Mathematics, 48*, 21–46.

Schoenfeld, A. (1994). *Mathematical thinking and problem solving*. Hillsdale, NJ: Erlbaum.

Stoffregen, T. (2003). Affordances as properties of the animal-environment system. *Ecological Psychology, 15*(2), 115–134.

Tikhomirov, O. (1981). The psychological consequences of computarization. In J. Wersht (Ed.), *Concept of activity in soviet psychology* (pp. 256–278). New York, NY: M.E. Sharpe.

Treffers, A. (1987). *Three dimensions, a model of goal and theory description in mathematics education*. Dordrecht, The Netherlands: Reidel.

Trouche, L., Drijvers, P., Gueudet, G., & Sacristan, A. I. (2013). Technology-driven developments and policy implications for mathematics education. In A. J. Bishop, M. A. Clements, C. Keitel, J. Kilpatrick, & F. K. S. Leung (Eds.), *Third international handbook of mathematics education* (pp. 753–789). New York, NY: Springer.

Van den Heuvel-Panhuizen, M. (2003). The didactical use of models in realistic mathematics education: An example from a longitudinal trajectory on percentage. *Educational Studies in Mathematics, 54*, 9–35.

Villarreal, M., & Borba, M. (2010). Collectives of humans-with-media in mathematics education: Notebooks, blackboards, calculators, computers and… notebooks throughout 100 years of ICMI. *ZDM, 42*(1), 49–62.

Watson, A., Jones, K., & Pratt, D. (2013). *Key ideas in teaching mathematics: Research-based guidance for ages 9-19*. Oxford, UK: Oxford University Press.

Wertsch, J. (1991). *Voices of the mind: A sociocultural approach to mediated action*. London, UK: Harvester Wheatsheaf.

Chapter 6
Digitally Expressing Algebraic Thinking in Quantity Variation

Abstract In this chapter, we describe and analyse a number of examples of 7th and 8th graders showing diverse ways of expressing their mathematical thinking in solving algebraic word problems with a spreadsheet. Different youngsters' approaches to situations where quantity variation is involved are characterised. The problems require finding an unknown value under a set of conditions that frame a problem situation. The use of the spreadsheet is thoroughly examined with the aim of highlighting the nature of problem-solving and expressing in the digital tool context as compared to the formal algebraic method; moreover, the ways in which students take advantage of the tool (being guided by and also guiding the spreadsheet distinctive forms of organising and performing variation in columns and cells) are important indicators of their algebraic thinking within the problem-solving activity. Finally, we pay attention to indicators of "co-action" in students' work on the spreadsheet as it tends to be more related to structuring solutions by means of creating variable-columns than with tentative ways of generating inputs in recipient cells.

Two data sources are used: the online competition collection of participants' solutions and in-class observations of 8th graders working on the same problems. Our analytic approach is a parallel analysis of solutions from the two settings, assuming that both data sources illuminate each other's differences and similarities. Our overall purpose is to acquire a view of youngsters solving and expressing algebraic problems concerning quantity variation in terms of their various uses of the spreadsheet.

Keywords Technology • Problem-solving • Algebra • Algebraic thinking • Variable • Variable-column • Spreadsheet • Co-action

6.1 Main Theoretical Ideas

The electronic spreadsheet, despite being a tool widely used in financial and management areas, has also proved to be an educational resource with great potential both in mathematical problem-solving and in the study of algebraic topics.

Hegedus (2013) reinforces the idea that technological affordances must become mathematical affordances and argues that meaningful integration of new learning environments can be developed through mathematisation of technological affordances. In the same study, Hegedus presents a set of future design principles (executable representations, co-action, navigation, manipulation and interaction, variance/invariance, mathematically meaningful shape and attributes, magnetism, pulse/vibration, construction and aggregation) and anticipates that these could provide innovative insights for researchers in the future. The idea of co-action, one of the features on this list, is an aspect to which we devote attention as we argue it is particularly pertinent to the study of algebra problem-solving with a spreadsheet.

6.1.1 Digital Representations in the Spreadsheet

The spreadsheet allows integrating different types of representations, such as natural language, through text, input of numbers and formulas and the creation of graphs, besides allowing objects produced by other technologies, for example, an image editor, to be included.

One feature that distinguishes the spreadsheet from other computing environments is the fact of supporting the connection between different registers (numeric, relational and graphical). The selection and dragging of the "fill" handle of a cell (or set of cells) is an operation that allows the generation of numerical linear sequences. These sequences can have different characteristics according to their origin and may have null increase, constant increase or others, according to the selected numbers or the formulas introduced and replicated. Such possibility of purposefully filling down a column is a way of creating what may be called a *variable-column*.

When handling a spreadsheet, students have the opportunity to discover and understand the meaning of a cell, a column and a formula and what it means to drag down the handle of a cell with a formula, as they automatically receive numerical feedback returned by the computer. This kind of vocabulary, which is characteristic of the spreadsheet, is far from the language commonly used in mathematics—"the vocabulary in the spreadsheet is far from the mathematical one, the user must even *create* it by himself [*sic*], there is no official reference to help him [*sic*]" (Haspekian, 2003, p. 123). According to Haspekian (2005), "communicating with a spreadsheet requires that pupils use an interactive algebra-like language, which focuses their attention on a rigorous syntax. This is why it is said that spreadsheets help to translate a problem by means of an algebraic code" (p. 113).

We claim that in the case of algebra problems, the spreadsheet can help students find and express relationships between variables in a given problem. In addition, it provides controlled means based on instantaneous and constant numerical feedback, which allows experiments to support establishing conjectures and may even help finding possible mistakes.

The numbers in the cells of the spreadsheet can have different natures. A number can be a numeric input, an output of a formula or an output of a linear number sequence with a given increase, automatically generated by the spreadsheet. In the case where the number is an output of a formula, the appearance of the current

6.1 Main Theoretical Ideas

cell is just a number. However, the cell may temporarily show the appearance of a formula—which occurs when the formula is introduced or thereafter, when the cursor is placed over the cell and one observes the formula bar. Thus, an important feature of the worksheet is to cover the formulas (i.e. the algebraic part) while maintaining the numeric always visible (Haspekian, 2003).

In solving a mathematical problem, especially under the perspective of the unity of solving and expressing (see Chap. 4), the capability to record and organise information and the clarity in expressing ideas and in building a solid argument are important capabilities. Representations are a fundamental means to contribute to such clarity and to the expression of mathematical knowledge. However, the systems of representation can be transparent or opaque. This distinction, made some time ago by Lesh, Behr, and Post (1987), means that the representations may be closer to the ideas they are intended to illustrate or be further apart when stressing just some aspects of these ideas while fading others. This transparency/opacity of representational systems has been expanded by Zazkis and Liljedahl (2004) who have suggested that there is a certain degree of opacity in any mathematical representation. In the case of the representational register of the spreadsheet, the first contact with it in the educational environment suggests a large opacity, which nevertheless tends to dissipate as students gain either familiarity with the specific language of the spreadsheet or greater agility in keeping a connection between algebraic thinking and the operations performed with and by the spreadsheet.

Figure 6.1 shows the spreadsheet table for the solution of a problem about finding the ages of three siblings, under conditions that relate them, where a student (Monica; all names are pseudonyms) enters the numbers manually without using formulas. This solution shows that there is a large opacity in the student's spreadsheet representations for solving the problem. On the other hand, in João's spreadsheet solution, shown in Fig. 6.2, the choice of the independent variable as well as the establishing of the given relationships is visible. The representations created by João show a lower degree of opacity than Monica's.

Carlos	Ana	Ricardo	Produto das idades dos irmãos *(Product of the ages of the brothers)*	Quadrado da idade de Ana *(Square of the age of Ana)*
0	1	2	0	1
1	2	3	3	4
2	3	4	8	9
3	4	5	15	16
4	5	6	24	25

Fig. 6.1 Monica's solution in the spreadsheet

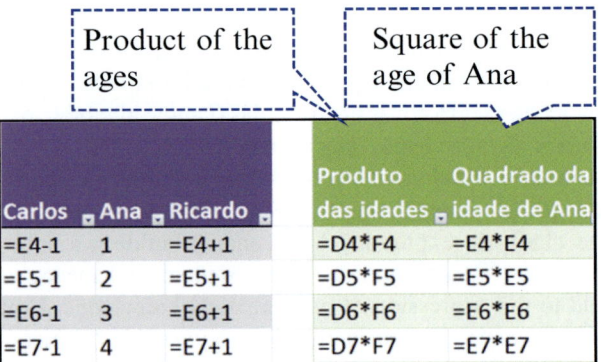

Fig. 6.2 João's solution in the spreadsheet

6.1.2 Algebraic Thinking

According to Kieran (2004), the work in algebra can be divided into three areas according to the nature of the mathematical activities involved: generational, transformational and global/metalevel activities. Generational activities correspond to the construction and interpretation of algebraic objects. Transformational activities include simplifying algebraic expressions, solving equations and inequalities and manipulating expressions. Finally, global/metalevel activities involve problem-solving and mathematical modelling, including pattern generalisation and analysis of variation.

The nature of algebraic thinking depends on the age and mathematical experience of the students. Students at a more advanced level may naturally use symbolic expressions and equations instead of numbers and operations. However, for students who have not yet learnt algebraic symbolic language, the more general ways of thinking about numbers, operations and notations may, in effect, be considered algebraic (Kieran, 2007). Contexts that involve numbers, functional relationships, number patterns and regularities and other properties are an essential foundation for the understanding of algebraic structures. For instance, writing symbolic numerical relations may favour the use of letters. However, the use of technological tools allows other representations for such relations, as well as new forms of exploration of such relations, which may be seen as analogous to generational and transformational activities in algebra. Thus, it seems appropriate that such new representations, and the mathematical thinking associated with them, are included in the field of algebra (Kieran, 1996). Moreover, Lins and Kaput (2004) claim that algebra can be treated from the arithmetic field stance, since there are many properties, structures and relationships that are common to these two areas. Therefore, arithmetic and algebra may be developed as an integrated field of knowledge.

According to Zazkis and Liljedahl (2002), the term algebra encompasses two distinct aspects, algebraic thinking and algebraic symbolism, stressing that the presence of algebraic symbolism should be taken as an indicator but that the absence of algebraic notation should not be judged as an inability to think algebraically. This idea is in the spirit advocated by Radford (2000) whereby students are able to think algebraically even when they do not resort to algebraic symbolism in their written productions.

In our research, we adopt this perspective, considering algebraic thinking as a broad way of thinking that is not limited to the formal procedures of algebra. This entails separating algebraic thinking from algebraic symbolism (Kieran, 2007; Zazkis & Liljedahl, 2002).

6.1.3 Problem-Solving with the Spreadsheet and the Development of Algebraic Thinking

The spreadsheet has proved to be a powerful tool in mathematical problem-solving and particularly in the development of algebraic thinking through problem-solving activities as highlighted by several authors (e.g. Ainley, Bills, & Wilson, 2004; Dettori, Garuti, & Lemut, 2001; Rojano, 2002). One of the gains of connecting algebraic thinking and the use of spreadsheets is the creation of a significant environment to induce students into an algebraic language that facilitates the construction of algebraic concepts, especially in what concerns working with functional relations, sequences and recursive procedures.

Using the spreadsheet in the context of problem-solving emphasises the need to identify the relevant variables involved in a problem situation and fosters the search for variables that depend on other variables, resulting in composites of relations between variables. The definition of intermediate relations, by means of spreadsheet formulas in intermediate dependent columns, meaning the decomposition of more complex relations in chained simpler ones, is a special feature inherent to the use of the spreadsheet that amounts to important results in solving algebraic contextual problems (Carreira, 1992; Haspekian, 2005). Moreover, as noted by Haspekian (2005), a spreadsheet also allows an algebraic organisation of apparently arithmetical solutions, and this kind of hybridism, where arithmetic and algebra naturally cohabit, becomes an educational option that may help students in moving from arithmetic to algebra (Kieran, 1996). Spreadsheets can act as a bridge between arithmetic and algebra by helping students generalise patterns, develop an understanding of variable, facilitate transformation of algebraic expressions and provide a space to explore equations (Tabach, Hershkowitz, & Arcavi, 2008). In addition, spreadsheets support students to focus on the mathematical reasoning by freeing them from the burden of calculations and algebraic manipulations (Ozgun-Koca, 2000).

However, as Dettori et al. (2001) have noticed from their research on 13–14-year-old students' work with spreadsheets on algebraic problems, "spreadsheets can start the journey of learning algebra, but do not have the tools to complete it. Being able to write down parts of the relations among the considered objects, but not to synthesize and manipulate the complete relations, is like knowing the words and phrases of a language, but being unable to compose them into complete sentences" (p. 206). What still remains a research issue is to understand the scope of the spreadsheet contribution in going further than just the recognition and manipulation of relations among objects to a broader understanding of the algebraic foundations of the methods for solving algebraic conditions.

6.1.4 Expressing Algebraic Thinking and Co-action with the Spreadsheet

Co-action is an idea proposed by Moreno-Armella, Hegedus, and Kaput (2008) to explain and describe the changes that the use of digital technology brings into students' mathematical activity. This idea of co-action is related to the fact that students are at the same time guiding and being guided by the dynamic and interactive digital environments, like the spreadsheet. Moreno-Armella and Hegedus (2009) state it as follows: "the student and the medium re-act to each other and the iteration of this process is what we call co-action between the student and the medium" (p. 510). The productions are thus the result of a collaborative work between students and the media provided by the technological tool.

With a spreadsheet, like with other tools of a dynamic and interactive nature, such as the case of GeoGebra, through a simple command, there is access to how the subject interacted with the tool. In particular, there is data on how the relationships between the cells, columns or rows were conceived and what is the computational process translated by the formulas introduced (MS Excel allows the user to switch between displaying formulas and their values on a worksheet). In the case of defined intermediate relations, there is access to the order in which the relations were created. This feature of the spreadsheet is important because in the analysis of a worksheet whose construction has not been witnessed, there may be evidence of the type of co-action that may have existed during the activity between the subject and the tool.

The spreadsheet provides a learning environment in which problem-solving can be explored in a dynamic way, when compared with pencil and paper. The transition from inserting symbols in a static environment to interacting with a dynamic environment brings out new forms of symbolic thinking. In this regard, Moreno-Armella et al. (2008) consider five distinct stages of computational development: the first two relate to stages of static non-computational environments; the third corresponds to a stage in which the representations are static but originating from a computational environment, such as a calculator; the fourth stage is called discrete dynamic, having the example of the spreadsheet where you can create a list or a chart and act on them interactively and where action can coexist; and the fifth and last stage is called continuous dynamic and is based on the above, where computational environments are very sensitive to the actions of the user, letting you drag or move objects, look into their mathematical properties and permanently reorient your perspectives about what is happening.

The work with the spreadsheet transforms the nature of the interactions that students have with mathematical representations to the extent that those become encapsulated in a medium with specific characteristics. The solution of a problem in the spreadsheet arises from the student's ongoing collaboration with the tool; both the student and the spreadsheet act and react to each other throughout the activity (Moreno-Armella & Hegedus, 2009). This type of work has significant consequences for the expression of students' mathematical thinking, particularly of algebraic thinking, during their problem-solving and expressing.

In solving a problem with the spreadsheet, the co-action between the student and the tool begins with the need for structuring the conditions of the problem in columns or cells that are assigned particular roles. This procedure allows connecting a set of numbers (e.g. in a column) with a single name (or column heading) which is consistent with an idea of variable, and it is an action that pushes students' reflection and helps them to understand the mathematical meaning of the relations among variables (Wilson, 2007). The introduction of numerical data in different cells, which may or may not include the use of formulas, becomes part of establishing the relationships described in the problem situation. Furthermore, students can analyse the immediate feedback provided by the spreadsheet and redirect their actions in a permanent flow of interactions with the spreadsheet. This work, based on the identification and materialisation of functional relationships, induces an algebraic organisation in the way of addressing the problem (the creation of a conceptual model of the situation) that apparently has a numerical look (Haspekian, 2005). Students are then able to inspect their table to get the solution supported by the results shown in the spreadsheet.

Problem-solving with spreadsheet and the co-action involved provide a stimulating working environment that fosters a greater understanding of the relationships of dependence between variables and encourages students to submit solutions gradually more algebraic rather than purely arithmetic (Rojano, 2002).

6.2 Context and Method

In the remainder of this chapter, we describe and analyse how students from 7th and 8th grades (12–14-year-olds) express their mathematical thinking in solving problems with the spreadsheet. We set out to know more about students' digital representations in solving quantity variation problems, in relation to their algebraic thinking and their problem-driven algebraic models; another issue that we expect to be related to the previous ones brings us to the attempt of unveiling the co-action between the student and the spreadsheet while solving a problem.

The data analysed refer to the solutions given to two algebraic word problems, involving whole quantity variation, proposed in the Qualifying phase of the competition SUB14 (see Chap. 1). Besides considering the data provided by all the solutions submitted by students who were engaged in the online competition, the same two problems were also given to a class of 8th graders as part of their periods of supervised study (non-curricular classes) in a public middle school located in the south of Portugal. Such class periods took place, from January to May, during the school year 2009–2010, on a regular basis, once a week, for 90 min.

Solving the two selected problems by a formal algebraic approach (namely using inequality solving techniques and systems of simultaneous linear equations) was beyond the reach of the students' school grades in terms of the topics comprised in the official school mathematics curriculum. Therefore, the use of the spreadsheet in the classroom was expected to provide the chance to see which avenues would be opened by this particular technological tool in students' development of their initial conceptual models and in their ways of expressing such models with this digital medium.

In the empirical context of the 8th grade classroom, five students were selected for this study: two of them were working as a pair and the other three individually. The students were given the freedom to choose whether they wanted to work in groups or individually as part of the didactical contract established in the classroom. In both cases, the teacher frequently engaged in dialogue with the students and asked questions whenever necessary to appreciate students' reasoning and approaches. They had previously obtained some experience in solving word problems with a spreadsheet in their regular mathematics classroom, from which they acquired the basics of the spreadsheet functioning.

Many of the problems that were explored with this class were chosen from the ones proposed at the competition SUB14. The problems were solved in school, and afterwards the students could send their answers to the competition, if they wished so. This was consistent with what the competition promoted: allow and encourage discussion of the problems with parents, teachers and fellow participants. The teacher of this class felt that the problems of the competition were important for students' mathematical learning, and she decided to use them in supervised study periods as it would mean more time to work with the computer and to discuss the problems.

The possibility of participants sending their answers to the competition in different digital formats (including spreadsheet files) was seen as an inducement to engage students in working on word problems with the use of the spreadsheet and an opportunity to develop students' algebraic thinking.

In the classroom, the detailed recording of students' processes was achieved with the use of *Camtasia Studio*. This software allows the simultaneous collecting of the students' dialogues and the sequence of computer screens that show all the actions that were performed on the computer. Thus, we were able to analyse students' conversations while we observed their operations on the spreadsheet. This type of computer protocol is very powerful as it allows the description of the user's actions in real time on the computer (Weigand & Weller, 2001).

The decision to collect data from a classroom setting in addition to the solutions obtained online in the competition SUB14 relates to the fact that many mathematics teachers see the competition as a potential resource for mathematics learning in their classes (see Chap. 3). We think that such live data can provide evidence of how the process of problem-solving and expressing can also happen in out-of-school environments. Moreover, the data serve the purpose of looking at possible benefits of spreadsheet use for the expression of mathematical reasoning and at the ways in which it may enhance the developing of algebraic thinking without demanding the use of algebraic symbolism and formal algebra techniques.

After collecting all the digital solutions submitted to the SUB14 competition containing Excel files, we analysed how students used the spreadsheet to solve the problem: how the columns were generated, how the relations were set (with or without the use of formulas) and the diversity of complementary representations used. From this analysis, we found typical solutions (or apparent categories). After this categorisation, we analysed the solutions produced in the classroom with the use of Excel and tried to establish a correspondence between solutions in the same category from the two sources. In the case of the solutions obtained in the classroom, we also looked through the audio and computer screen recordings in order to track students' actions during their work with the computer. If we found solutions from the classroom that were different from those coming from the competition, they were subject to further categorisation.

In presenting the results, we also look to establish a hierarchy of solutions in terms of the generality of the conceptual model involved and its closeness to the algebraic formal language.

6.3 Data Analysis

6.3.1 *The First Problem: The Treasure of King Edgar*

The first problem contains several conditions that relate to each other, and the statements "gets more… than" and "receives fewer… than" involve an element of ambiguity and make the problem complex, for understanding it, for translating into algebraic language and for solving it (Fig. 6.3).

King Edgar of Zirtuania decided to divide his treasure of a thousand gold bars among his four sons. The royal verdict is:
1 - The 1st son gets twice the bars of the 2nd son.
2 - The 3rd son gets more bars than the first two sons together.
3 - The 4th son receives fewer bars than the 2nd son.
What is the largest number of gold bars that the 4th son may receive?

Do not forget to explain your problem solving process!

Fig. 6.3 Problem #2 from the SUB14 competition (edition 2009–2010)

A possible algebraic approach to the problem is presented in Fig. 6.4.

Among the 276 participants in the SUB14 competition that sent correct answers to the problem, only nine made use of the spreadsheet. In the classroom, we obtained ten solutions to the problem produced with the spreadsheet.

We now analyse four solutions in detail, two from each environment—the competition and the classroom—that are representative of all the solutions that were produced by resorting to the spreadsheet.

We start by presenting one solution outlined by a group of three students in 7th grade, Abel, Bruno and Carlos, who were participating in the competition. The group made use of the spreadsheet to prepare a table and organise their trials to obtain the answer to the problem. Students named four columns, one for each son and two other columns that were defined to compute the sum of bars given to the four sons and to compute the difference between that sum and the total of 1000 bars, as seen in Fig. 6.5.

Fig. 6.4 A possible symbolic algebraic approach to the problem

$$\begin{cases} s_1 + s_2 + s_3 + s_4 = 1000 \\ s_1 = 2s_2 \\ s_3 > s_1 + s_2 \\ s_4 < s_2 \end{cases} \quad (\max s_4)$$

s_i – number of bars of son i $(i = 1\ldots 4)$

1st son Primeiro Filho	2nd son Segundo filho	3rd son Terceiro filho	4th son Quarto filho				Difference	
100	50	151	49	total:	350	Diferença:	650
150	75	226	74	total:	525	Diferença:	475
200	100	301	99	total:	700	Diferença:	300
250	125	376	124	total:	875	Diferença:	125
260	130	391	129	total:	910	Diferença:	90
270	135	406	134	total:	945	Diferença:	55
272	136	409	135	total:	952	Diferença:	48
276	138	415	137	total:	966	Diferença:	34
280	140	421	139	total:	980	Diferença:	20
282	141	424	140	total:	987	Diferença:	13
284	142	427	141	total:	994	Diferença:	6
286	143	430	141	total:	1000	Diferença:	0

Fig. 6.5 Excerpt of Abel, Bruno and Carlos's solution (participants in the competition)

6.3 Data Analysis

Along with the table, students provided a rationale of their own procedures written in natural language in the spreadsheet cells:

> We carried out a table in which the first son, as said in the problem, has twice the bars of the second son. The third child (in our table) has just one more bar than the sum of the first and second sons' so that there are more gold bars left for the fourth child. The fourth son has always (in our table) one bar less than the second son so that he will get the largest possible number of gold bars. Initially we chose the number 100 as a basis to start our table regardless of the outcome for the total of bars is 1000 or not. After we had our first "trial" we started making adjustments always increasing by 50 the number of gold bars of the first son. But we always calculated the difference so that when we would approach the sum of 1000 we would change the increase by 50 to an increase by 10 or by 2, depending on whether being closer to 1000 or not. In the last row of our table we kept the 141 gold bars for the fourth child because we could not further increase this number and still meet the problem conditions. Therefore in order to have no bars being left (the 6 remaining bars), we "gave" two more bars to the first son, "gave" one more bar to the second (for the second son must have half of the bars of the first son) and we "gave" the three remaining bars to the third son.

The students' expression of the solution includes an explanation of how they built the columns and of their decision making, such as the distribution of the six remaining bars among the sons. This group uses the organisation, in columns, afforded by the spreadsheet to make explicit the conditions given in the problem. Their strategy was to use trial and refinement to obtain successive approximations to the solution.

In considering the way they use the spreadsheet, there is a certain degree of opacity in the representations provided in the spreadsheet, since they did not resort to formulas, which would have facilitated the calculations and the nearly automatic filling of the table itself, as it is illustrated in Fig. 6.6.

However, as seen in the figure, we can understand the relationship between students' work on the spreadsheet and the conditions of the problem statement written in algebraic language. The fact that students solved the problem without resorting to

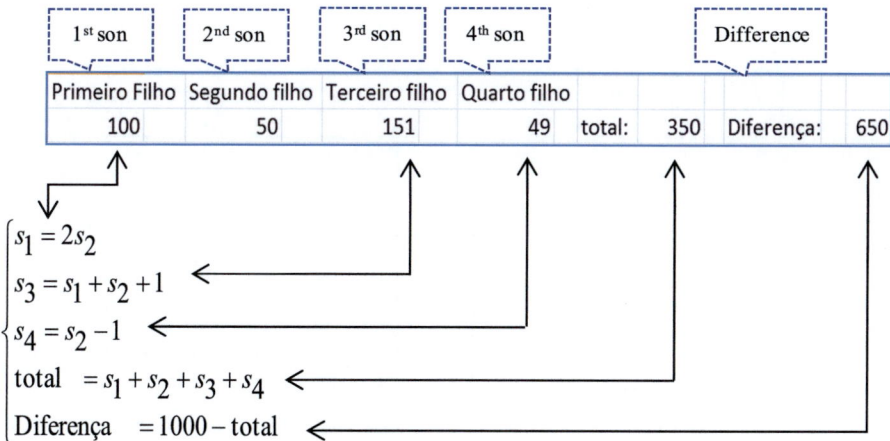

Fig. 6.6 A translation of the group's model on the spreadsheet into symbolic algebraic language

formulas did not prevent them from correctly establishing the relationships between the variables in the problem, by means of the spreadsheet columns. Their solution, which includes the table and the explanation in natural language, allows us to understand what their conceptual model of the situation was and how it expressed their mathematical thinking, in particular the establishment of the algebraic relationships and the strategy (trial/refinement) to achieve the solution.

Next, we present a solution which was developed in the classroom and is close to the one given by the previous group of participants in the competition.

Marcelo is one of the students in the class who has worked individually. After reading the problem, he began by introducing manually in column A of the spreadsheet the multiples of 100 up to 1000, but he never went back to using this column. Next, Marcelo assigned and named a column to each of the four sons and a fifth column to the total of gold bars distributed. In that last column, he entered a formula to compute the sum of gold bars and used it as a way to check his mental calculations (Fig. 6.7). Then, row after row, he started writing values in the cells corresponding to the four sons. In doing this, he used two different strategies for assigning values to the four sons. Sometimes he worked in the following order: second, first, fourth and third. The input of the values went on as follows: choosing a value for the second son and then mentally doubling it for the first son and then subtracting one unit to the second son's number of bars to get the value for the fourth son; add the three values corresponding to the second, first and fourth sons and calculate the difference to 1000 in order to find the third son's number of bars. Otherwise, he used a different order: second, first, third and fourth. He started choosing a value for the second son and then doubling it for the first son and then adding the first two and increasing the result by one unit to get the third son's number of bars; add the number of bars of the three first sons and calculate the difference to 1000 in order to know the fourth son's number.

Marcelo did not display the relationships between the numbers of bars of the four sons—using formulas or otherwise—but he kept them always present in his thinking.

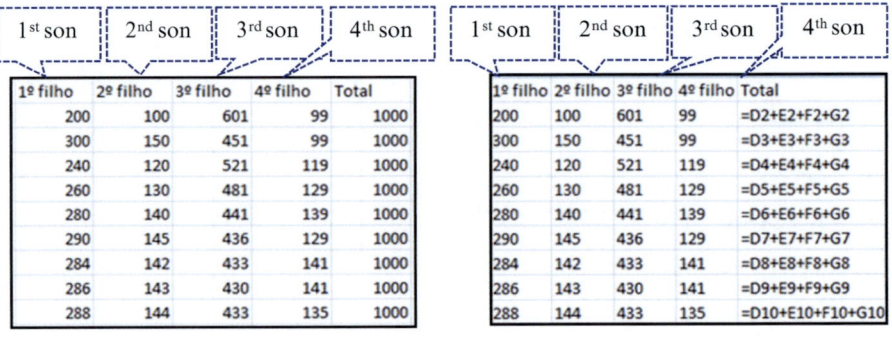

Fig. 6.7 Excerpt from Marcelo's solution (student from the classroom)

6.3 Data Analysis

The task required a great effort for the student, since in each attempt he had to recall the conditions to observe while carrying out the calculations mentally.

During his work, and after several tests, he called the teacher.

> Marcelo: Teacher, I found the best! [The value 139 was obtained in a cell of the column assigned to the fourth son]. If I choose 150 bars [in the column of the second son] it won't do. I've tried it.
> Teacher: But is it the maximum number of bars that the fourth may get? Is it?
> Marcelo: I went from 100 to 150 [for the second son], and it turns out that 150 gets worse because the other gets over 450 but the last one falls to 99.

The teacher suggested Marcelo do some more experiments to which he replied that he had already made some, for example, with the values 160 and 170 for the second son. So she made another suggestion:

> Teacher: Here you already got an excellent value, and it increased significantly from 130 to 140 [referring to the column of the second son]. So, try around these values.
> Marcelo: I'll do it with 145.
> Teacher: Did it turn out any better?

The student continued to make trials, but he took a while to do the calculations mentally, and the teacher asked:

> Teacher: But why don't you make Excel to do the computations?

The student did not seem to know what to answer, probably indicating that his mastery of the tool was not strong enough. So he continued to make trials, always doing the calculations mentally. Finally, he found the solution 141 bars, confirming that it was the highest possible number of bars for the fourth son. In his answer to the problem, the student wrote: "I solved this problem taking into account the conditions of the problem, making four columns, one for each son, and trying to find a higher number for the 4th".

We claim that Marcelo has developed algebraic thinking by focusing on dependence relationships between different variables to find the optimal solution. As the student stated, he took into account the five conditions of the problem and expressed them in the spreadsheet columns. From the standpoint of an algebraic approach, the student began by choosing an independent variable (the second son's number of bars) and established relationships to express the number of bars for each of the other sons.

The diagram in Fig. 6.8 summarises the symbolic translation of the student's algebraic thinking in solving the problem and shows how the spreadsheet allowed dealing with several simultaneous conditions and manipulating them, by means of numbers rather than letters and symbolic algebra. It is important to note that the student understands that there were two different ways to try to maximise the number of bars of the fourth son: one way was to assign the fourth son just a bar less than the second son; the other way was to assign the third son just one bar over the sum of the first and second sons' bars. The student alternates between these two strategies in making trials to find the highest value for the fourth son's bars. The fact of using the two strategies led him to find two different solutions with the same maximum number of bars for the fourth son.

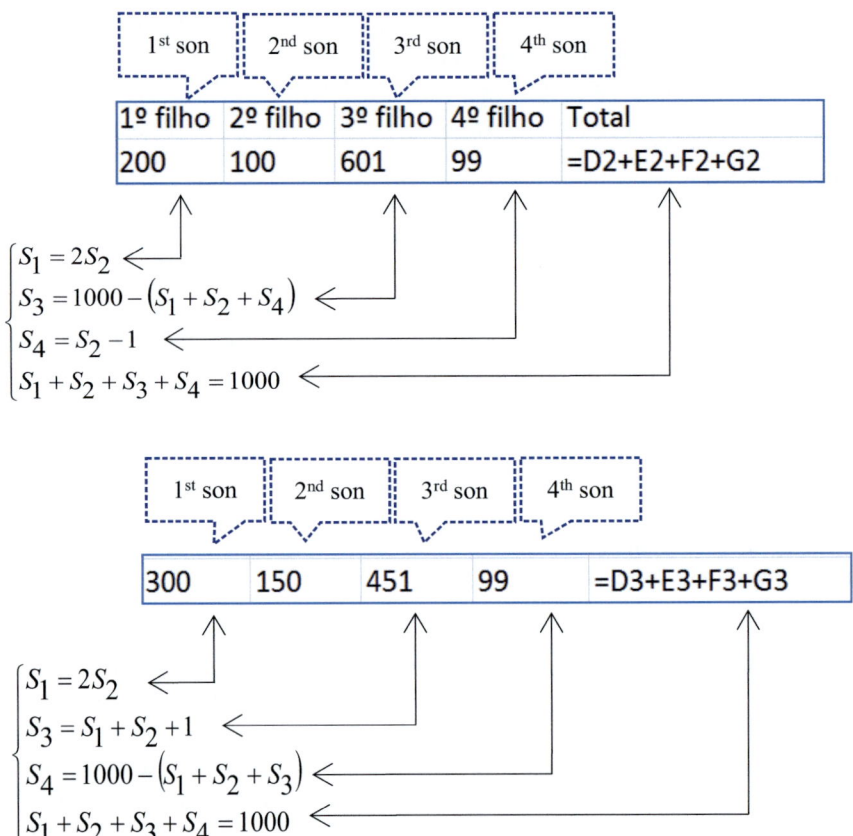

Fig. 6.8 A translation of Marcelo's model on the spreadsheet into symbolic algebraic language

The capture of the computer screens and student's speech allows us to reconstruct Marcelo's activity and to corroborate that his problem-solving was the product of a weak co-action between him and the spreadsheet. The student performs an intense computational work (in his trials and checking), whereas the spreadsheet only provides a control over the total of bars and also a way of organising the variables. In a sense, the spreadsheet is used in a static way, similar to an enhanced calculator, suggesting a use of the tool corresponding to the characteristics of the third stage of computational development described by Moreno-Armella et al. (2008).

The solutions presented above (one from the online competition and one from a classroom) have some similarities but also some distinctions. The similarities concern the construction of the two initial columns; the distinctions are related with the way the conditions in the problem are represented and explored on the spreadsheet,

6.3 Data Analysis

including the way in which the column for the total of bars is used (as a variable or a constant).

Another answer from the classroom that we examine is that of Ana's, who like other classmates, undertook the construction of a table. On the spreadsheet, she started to write the column headings, "gold bars" and then "first son", "second son", "third son" and "fourth son", one in each column upper cell. In the first column, just under the heading "gold bars", she inserted one after the other the integers up to 9 and dragged the sequence down until the value 1000. This suggests that those columns were identified as the variables of the problem, which means that the student considered that the number of bars of gold could be a variable between 1 and 1000. However, the student never used this column again, probably by having realised that it would not work as an independent variable.

Then she inserted the values 2 and 1 on the first row of the columns referring to the first son and the second son (cells D4 and E4), respectively. Then she inputted the values 4 and 2 in the following row of the correspondent columns (D5 and E5) and the values 6 and 3, again in the next row of the same columns (D6 and E6). Thus, she established the beginning of a number pattern by a certain increment in each of the columns D and E. Next, she dragged the fill handle, extending the sequences until reaching the row corresponding to the value 1000 in the column used for the total of bars. In cell F4 (corresponding to the third son), she entered the formula "=D4+E4", and, with the cursor still on the cell, before completing the operation, she asked for the teacher's help. The following is a transcript of the dialogue that took place:

> Ana: And now, teacher, shall I do enter?
> Teacher: The third son gets more bars than the first two sons together… You may wonder how many more… But you must understand that the fourth son… [The teacher does not conclude the sentence and turns herself to address the whole class.]
> Teacher: You must not forget that we are looking for the highest possible number of bars that the fourth son can get.

Later, Ana called the teacher again as she apparently had come up with a different idea on the formula to use and the dialogue resumed.

> Ana: The third son gets more bars than these two. [She was pointing to the first and second sons' columns while writing the formula "=E4+D4+1" in cell F4.]
> Teacher: And now, what do you need to do?
> Ana: Drag it down.
> Teacher: And now?
> Ana: The fourth son will get fewer bars than the second one. So, if the second one starts with 1, here I have to start with 0.
> Teacher: And how are you going to make it? Would it be possible to create a formula?
> Ana: Yes, it's this one here [she was pointing to cell E4] minus 1.
> The student inserted the formula and then dragged the fill handle.
> Teacher: And now, what are you going to do next?
> Ana: I am going to add them all.

The student enters the formula "=G4+F4+E4+D4" in cell H4 and drags the fill handle, extending the column up to a certain value, as she almost immediately realised to have exceeded the limit of 1000 bars.

Ana: It already exceeds. [The total of bars had exceeded 1000.]
Another student: This gives me 1001.
Ana: Me too, it gives me 1001.

Ana called the teacher once again as she was unsure of the answer to the problem.

Ana: This is wrong, isn't it?
Teacher: There's a brother who is receiving one bar more than he should, isn't there? Which one can he be?
Ana: This one, the fourth brother... 142, then he will get 141!

The student pointed out the solution, coloured the cells and explained her reasoning directly on the spreadsheet cells. An excerpt of the table created by Ana with the spreadsheet can be seen in Fig. 6.9.

In the end, the student deleted all the table rows that contain unnecessary information (which far exceeded the total number of bars). She also gave a detailed explanation of her problem-solving process on the spreadsheet, as follows:

> The largest number of bars that the fourth child may receive is 141 gold bars. As the king decided to divide his treasure of a thousand gold bars among their children I decided to solve the problem in Excel. In the first column I have put the number of gold bars and dragged. Then I made two columns for the 1st and 2nd sons knowing that the relationship between them is that the 1st son gets twice the number of gold bars of the 2nd son. In the 3rd column it was the 3rd son and he was receiving more bars than the 1st and 2nd sons together, so I added the 1st and 2nd sons plus 1. The result was: 1st son = 286 bars, 2nd son = 143 bars and 3rd son = 430 bars. But what matters is the 4th son and as he gets fewer bars than the 2nd one, I did: (2nd son's bars) − 1, that equals to 0, and I dragged. Then I made the total and it showed 1001 but since the king had only 1000 bars I picked the result that came before that 142, which was 141. The maximum number of gold bars that the 4th child may receive is 141.

The data obtained from Ana's solving process show that either their language or their actions are steeped in the language and operation mode of Excel. The student knows that selecting and dragging some sequential numerical values in a column, or dragging a cell with a formula, produces a variable-column.

As we saw in the first two examples presented above, Ana also managed to express the problem in terms of relationships between columns in the spreadsheet (Fig. 6.10). However, Ana did so by using the automatic generation of sequences

Gold bars	1st son	2nd son	3rd son	4th son		Gold bars	1st son	2nd son	3rd son	4th son
barras de ouro	1º filho	2º filho	3º filho	4ºfilho	total	barras de ouro	1º filho	2º filho	3º filho	4ºfilho	total
1	2	1	4	0	7	1	2	1	=E4+D4+1	=E4-1	=G4+F4+E4+D4
2	4	2	7	1	14	2	4	2	=E5+D5+1	=E5-1	=G5+F5+E5+D5
3	6	3	10	2	21	3	6	3	=E6+D6+1	=E6-1	=G6+F6+E6+D6
4	8	4	13	3	28	4	8	4	=E7+D7+1	=E7-1	=G7+F7+E7+D7
141	282	141	424	140	987	141	282	141	=E144+D144+1	=E144-1	=G144+F144+E144+D144
142	284	142	427	141	994	142	284	142	=E145+D145+1	=E145-1	=G145+F145+E145+D145
143	286	143	430	142	1001	143	286	143	=E146+D146+1	=E146-1	=G146+F146+E146+D146
144	288	144	433	143	1008	144	288	144	=E147+D147+1	=E147-1	=G147+F147+E147+D147

xls file   xls file (command "show formulas")

Fig. 6.9 Excerpt from Ana's solution (student from the classroom)

6.3 Data Analysis

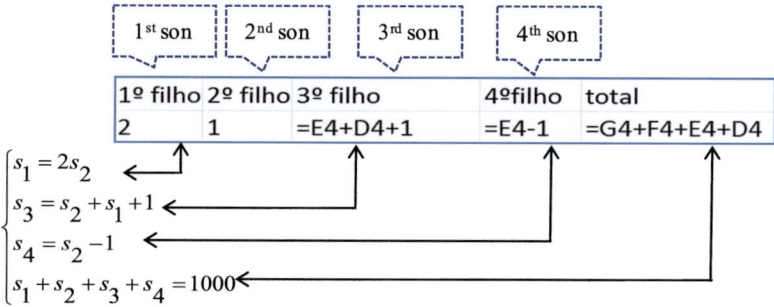

Fig. 6.10 A translation of Ana's model on the spreadsheet into algebraic language

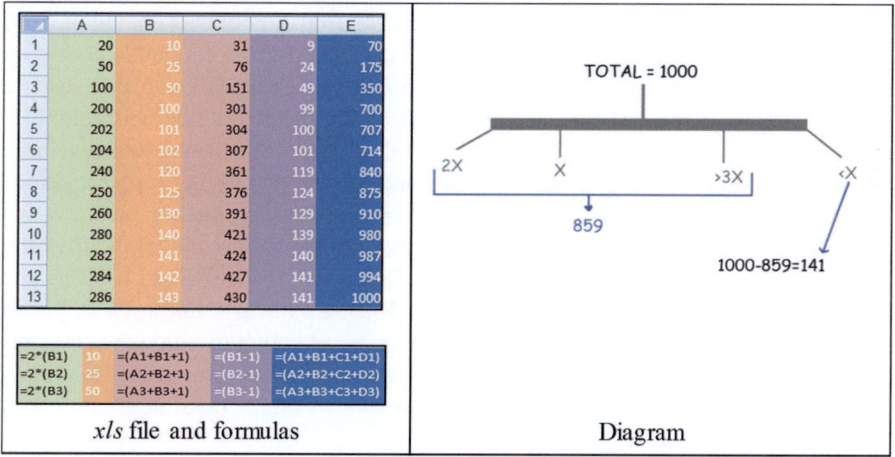

Fig. 6.11 Excerpt of David's solution (participant in the competition)

(by means of the *Autofill* feature of the spreadsheet) and inserting formulas, which allowed her to have her thinking more focused on the relations established than in the computations that she left to the computer.

We now turn to the analysis of David's solution, a participant from the 8th grade in the competition, who had enrolled individually. This student's answer to the problem contained a combination of representations with different characteristics and origins. He used a spreadsheet to prepare a table where he also featured a matching diagram showing some algebraic language, created with a drawing editing tool. These two pieces are clearly associated, as seen in Fig. 6.11 and as revealed in the student's explanation of his answer.

His explanation was presented in the following text sent as part of his e-mail message:

> Explanation/Reasoning: After reading the problem I thought of a way to solve it. I thought I would get it by trials, so I decided to do it. For this, I used Excel (attached is a diagram and an Excel table), using formulas as follows. The 2nd son would have X bars; the first would

have 2X, as the problem says that the 1st child gets twice the 2nd son's bars. The 3rd would have >3X (greater than 3X), as the problem says that the 3rd son will have more bars than the first two together and the fourth child would have <X (less than X), as the problem says that the 4th would have fewer bars than the 2nd son. By trial, I calculated until reaching the total of 1000 gold bars. I did not get to 1000, but I got to 1001, so I removed one bar to stay in 1000. At the end, the 4th son had 141 bars.

As he says, the student used the spreadsheet to make his trials to get the solution. Column B, referring to the second child's number of bars, is used as a container for putting values at his will; it is a column which works as the independent variable. The remaining columns refer to the number of bars of the other three sons and to the total of bars used. For its construction, the student used formulas expressing dependency relationships that allowed him to get, apparently with some ease, the required solution. In the scheme, the student displays with algebraic symbols the conditions of the problem, that is, the way in which the distribution of gold bars among the four sons is made. In his explanation, he shows to have established a connection between the spreadsheet language and the algebraic symbolic language. We can also observe that the order and roles of the columns in the spreadsheet coincide with the order and roles of the variables (independent and dependent) translated into algebraic language. This is revealing of how the student used the two digital representations in a complementary way to exhibit his model of the problem situation and how this model is showing a skeleton of a more standard mathematical model to solve the problem. We emphasise that the scheme and the type of symbolism used by the student are close to what would be a standard mathematical solution (with pencil and paper), while the solution produced in the spreadsheet retains unique characteristics, which could hardly be performed with pencil and paper, since there is a transformation of the algebraic variable into numbers, making it possible to stay with concrete numerical values.

Given the characteristics of his strategy of trial and refinement, the student succeeds in finding the two situations in which the fourth child will receive the maximum of 141 gold bars.

David is himself making an algebraic language translation of the work done in the spreadsheet, as can be seen in Fig. 6.12.

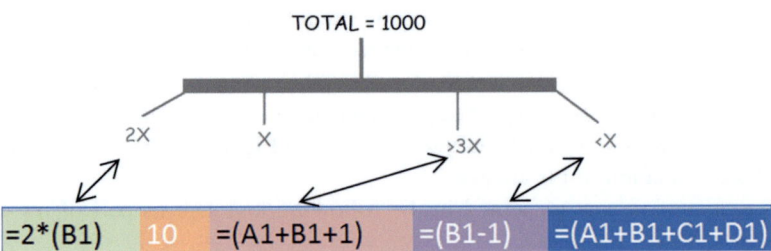

Fig. 6.12 David's translation into algebraic language of the relationships established in the spreadsheet

6.3 Data Analysis

We see that the student knew some of the features of the spreadsheet, like naming columns and writing formulas involving one or more cells, either input container cells or cells with values already derived from previously introduced formulas.

A key aspect to solve this problem is to make the number of bars assigned to the second son's change. In the four solutions presented, all students have adopted this strategy and appointed and selected either a variable-column or a column of container cells for the number of bars of the second son. This column operates in all cases, as the independent variable. The remaining columns are built up, with formulas or with direct inputs, expressing dependency relationships, some of which become composite relationships. In this way, students were able to observe the variation of the number of bars for each son, in particular for the fourth son, while checking the total number of bars assigned, and thus finding the answer to the problem.

It may however be noted that in the above first two and in the fourth solutions, a clear trial and refinement approach prevails, resulting from the fact that the cells of the column for the second son work as recipient cells where it is possible to test values and examine the outputs in the remaining cells. By contrast, in the third solution, the column for the second son works as a variable-column; that is, Excel is used to cycle through a sequence of values and to obtain the corresponding results on the remaining dependent columns.

In all the solutions, we can observe that the students chose to present, in addition to the table prepared in the spreadsheet, and as an integral part of the process of solving and expressing, an explanation in natural language and, in the latter case, even a diagram revealing a more symbolic algebra approach to the problem.

Revisiting the four solutions analysed, we can establish a hierarchy based on a progression from the use of recipient cells (numerical inputs) to the use of variable-columns (using formulas or the automatic generation of sequences). This ranking puts the solutions as the diagram in Fig. 6.13.

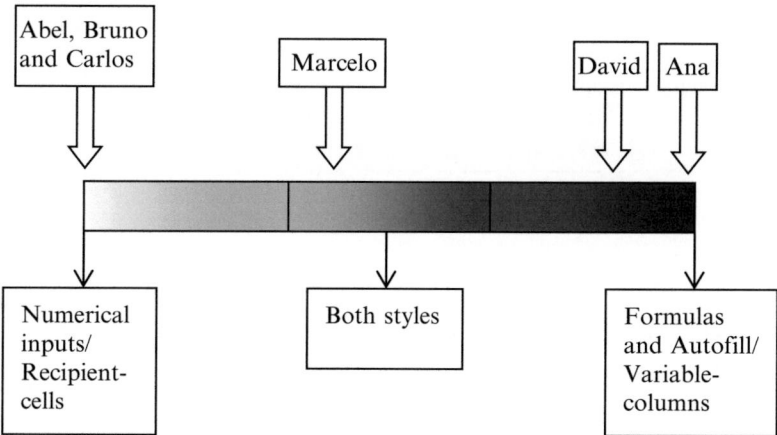

Fig. 6.13 A possible hierarchy of the analysed solutions

6.3.2 The Second Problem: The Opening of the Restaurant "Sombrero Style"

The second problem, presented in Fig. 6.14, has characteristics that were seen as interesting to be explored with the spreadsheet, namely, due to the fact that it may be solved by a numerical approach. To a certain extent, it proved to be different from other problems solved by the students in the classroom and in the competition. One of its features relates to the fact of being placed in the form of a narrative, which makes it pretty close to a real situation, where conditions are steeped in the story, thus requiring a very careful reading to identify those which are relevant to solving the problem.

A possible formal algebraic approach to the problem is presented in Fig. 6.15. Solving this problem through a formal algebraic approach, namely, using a system of three equations with three unknowns, such as presented, was beyond the curricular algebra content of 8th grade students.

In the competition, after a period of about 4 months from the start, the usual attrition shown in the Qualifying and the elimination of contestants had already decreased the number of participants. There were 42 correct solutions to problem #8, among which six were obtained with the spreadsheet. From the classroom, we got eight answers to the problem using the spreadsheet.

Gil, a participant in the competition from grade 8, sent his answer in two attached files, one prepared in Microsoft Word with the explanation of its procedures and reasoning and another in Microsoft Excel that he named as "Calculations".

The restaurant Sombrero Style was opened yesterday and I was there having dinner with three friends. The maximum capacity of customers – said the manager – is 100 people. Luckily I had booked a table for four, because when I got there several tables were already full with four people and one table had only three people. While I was waiting for the waiter to take us to our table, I counted the women and men who were in the restaurant and the number of women was exactly twice the number of men. What could have been the maximum number of people already in the restaurant when I came in?

Do not forget to explain your problem solving process!

Fig. 6.14 Problem #8 from the SUB14 (edition 2009–2010)

Fig. 6.15 Possible approach to the problem in algebraic language

w - number of women, m - number of men, y - total of persons, x - numbers of tables of 4 persons

$$\begin{cases} w = 2m \\ y = w + m \\ y = 4x + 3 \\ 0 < y \leq 100 \end{cases} (\max y)$$

6.3 Data Analysis

In his explanation, written in natural language, the student states:

To arrive at the answer to this problem I used the Excel file that I also sent attached. First I calculated the multiples of 4 and the multiples of 3 until about 100. Then I saw that the maximum number of people that could be there would be 96 because this was the total capacity of customers (100) minus the 4 people who have booked a table. Then I added 3 to all the multiples of 4, so that the table with three people mentioned in the problem was already added to all those multiples. From those numbers that I got, I began to look for those which were divisible by 3 (in numbers less than 96), since the number of women had to be twice that of men. The largest number that divided by 3 yielded an integer was 87. Therefore when that person who had booked a table entered the restaurant, there were already 87 people inside.

In this way, the student explained how the columns were devised in his spreadsheet and justified the reasoning that led him to find the solution (Fig. 6.16). In his spreadsheet table, no formulas were used.

In Fig. 6.17, we can see the correspondence between how Gil expressed the problem conditions on the spreadsheet and how they can be stated in algebraic language.

As can be seen, Gil found a simple way to address the problem and to get the solution easily, in particular by expressing the first two conditions of Fig. 6.15 in a single column of the spreadsheet: "mult 3" (multiples of three). So, this column was used in conjunction with the third column (multiples of four plus three).

Fig. 6.16 Excerpt from Gil's solution on the spreadsheet (participant in the competition)

mult 3	mult 4	mult 4+3
3	4	7
6	8	11
9	12	15
12	16	19
15	20	23
57	76	79
60	80	83
63	84	87
66	88	91
69	92	95
72	96	
75	100	
78	104	
81	108	
84	112	
87	116	
90	120	
93	124	
96	128	
99	132	
102	136	

Fig. 6.17 A translation of Gil's model on the spreadsheet into algebraic language

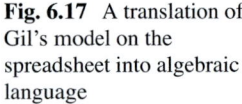
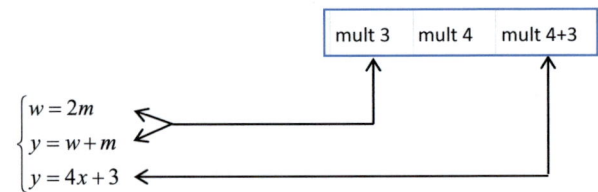

Returning to the classroom setting, when this problem was given, the students had already developed a considerable experience with the use of the spreadsheet to model relations within contextual algebraic problems. Therefore, some of them chose to use this tool to solve the given problem, engaging in the process of translating relations between variables and combining them in chained formulas.

However, the students expressed many difficulties in understanding the problem, particularly on the question about the maximum number of people seated in the restaurant before the last group of four people arrived. There were other obstacles that relate to the number of simultaneous conditions describing the distribution of persons by tables of three and four and also the division of clients by gender.

Maria and Jessica were working together in the classroom; they were one of the cases of students that chose to use the spreadsheet to solve the problem. Their spreadsheet table looked as shown in Fig. 6.18.

Maria and Jessica began by addressing the condition on the number of people seated at tables of four, as shown in the top of Fig. 6.18. In this process, the students identified the variable "number of tables of four" and in the next column generated the number of people sitting at these tables (multiples of four), using the Autofill feature to produce the sequence increasing by four. Then Jessica entered the constant-valued column referring to the three people which were seated at one of the tables (repeating the three in the following cells) and after created a column to compute the total number of people by using the formula "=H11+G11".

Afterwards, they separately represented the condition concerning the separation of clients by gender, as shown in the bottom of Fig. 6.18. The shaded row in each of the tables shows that the students sought to identify the solution by comparing the columns of totals in the two separate tables.

The work done by these students shows how the spreadsheet has helped them overcome the initial difficulties, in that it enabled them to work separately on the different conditions through independent tables and afterwards relate the feedback from each one to get the solution. The establishment of the first partial relations represented a transitional phase which facilitated the subsequent expression of the whole set of conditions stated in the problem.

In Fig. 6.19, we can see the correspondence that exists between Maria and Jessica's work on the spreadsheet and the problem conditions written in algebraic language. The students initially separated the conditions in two unconnected tables, but then, by inspecting the results for the total of people in both tables, they actually made the necessary connection in order to get the solution. Given the way they expressed the conditions in the spreadsheet, these students were even able to get

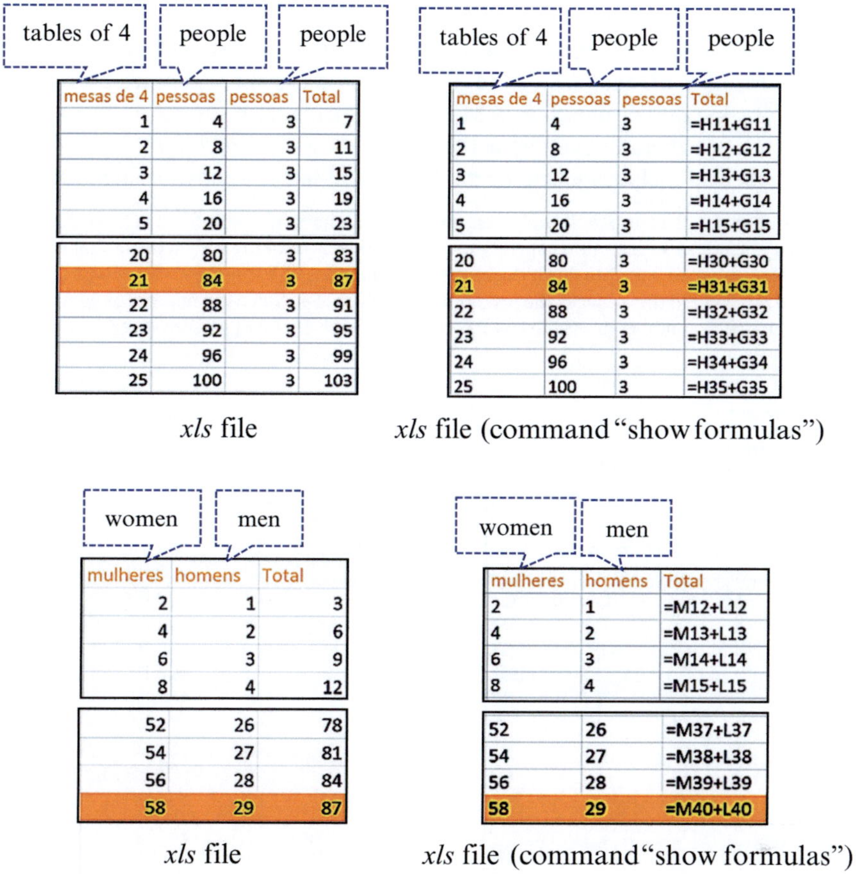

mesas de 4	pessoas	pessoas	Total
1	4	3	7
2	8	3	11
3	12	3	15
4	16	3	19
5	20	3	23
20	80	3	83
21	84	3	87
22	88	3	91
23	92	3	95
24	96	3	99
25	100	3	103

mesas de 4	pessoas	pessoas	Total
1	4	3	=H11+G11
2	8	3	=H12+G12
3	12	3	=H13+G13
4	16	3	=H14+G14
5	20	3	=H15+G15
20	80	3	=H30+G30
21	84	3	=H31+G31
22	88	3	=H32+G32
23	92	3	=H33+G33
24	96	3	=H34+G34
25	100	3	=H35+G35

mulheres	homens	Total
2	1	3
4	2	6
6	3	9
8	4	12
52	26	78
54	27	81
56	28	84
58	29	87

mulheres	homens	Total
2	1	=M12+L12
4	2	=M13+L13
6	3	=M14+L14
8	4	=M15+L15
52	26	=M37+L37
54	27	=M38+L38
56	28	=M39+L39
58	29	=M40+L40

Fig. 6.18 Excerpts from Maria and Jessica's solution in the spreadsheet (students from the classroom)

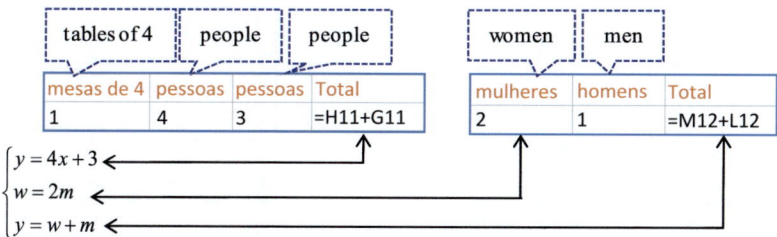

Fig. 6.19 Translation into algebraic language of the students' model on the spreadsheet

additional information beyond the strict answer to the problem (namely obtaining the number of women and the number of men already in the restaurant).

The following is an analysis of the production of Carolina, another classroom student, who, as with her classmates Maria and Jessica, also organised the conditions by separating them in two distinct tables, as shown in Fig. 6.20.

The student also includes a lengthy explanation of her problem-solving process:

> The maximum number of people who were in the restaurant when I got there is 87 people. Being the restaurant's maximum capacity of 100 persons, and taking into account that one of the tables only had 3 people this reduces the maximum possible number to 99. Given that I also had not entered this reduces the possibilities to 98 or less. Since the number of women is exactly twice the number of men, it can be concluded that the total of persons is represented

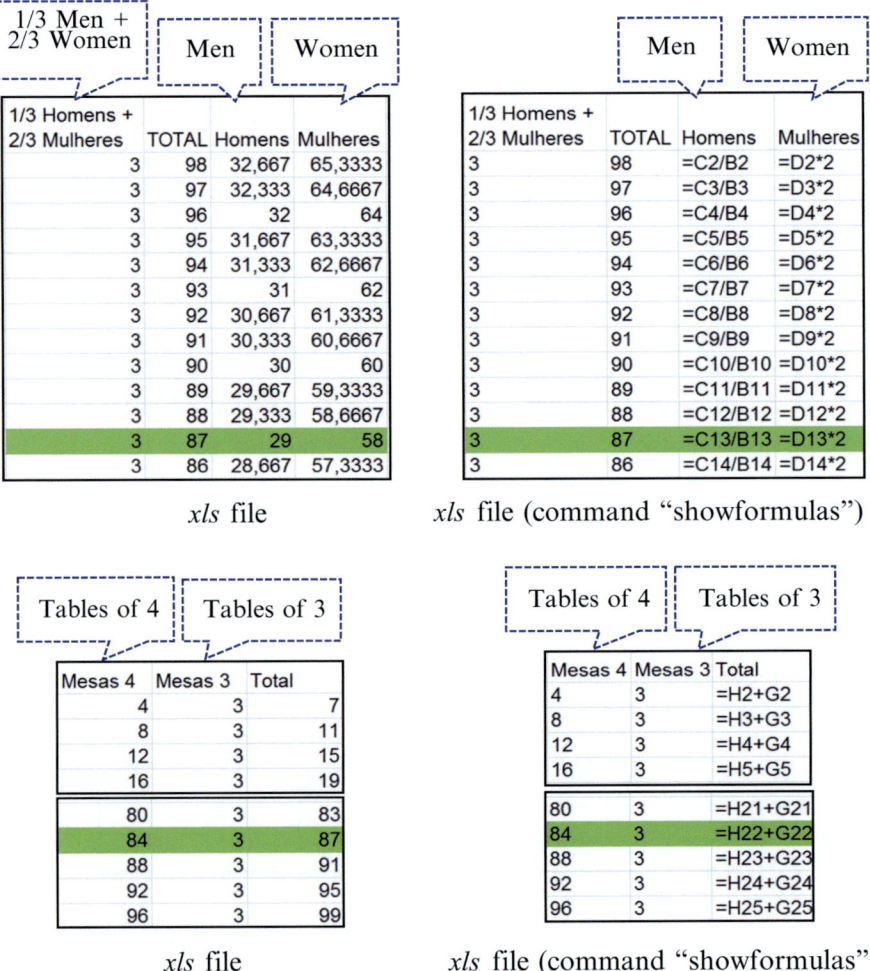

Fig. 6.20 Excerpt of Carolina's solution in the spreadsheet (classroom student)

6.3 Data Analysis

as 3-thirds, being one third of men, and two thirds of women. The 1st column thus refers to the 3-thirds mentioned. The 2nd column refers to the possibilities of the total of persons already in the restaurant, taking into account the reasoning explained above. The 3rd column with the formula "C2/B2" has to do with the possible totals divided by 3 to get one third, according to the reasoning explained above. The 4th column with the formula "D2*2" is about calculating two times the number of men, taking into account the reasoning explained above. In the final part, I have two columns for the tables of 3 and of 4 people. I have the multiples of four in a sequence, because there were several tables with 4 people. And as there was only one table with 3 persons, then the number three just repeats. Then finally I have the total of people sitting in tables, which brings us to the answer... So we have to return to the previous total number of people, and find an integer that matches a total number of people sitting at the tables.

Carolina begins her reasoning by making the separation of customers by gender. In a column, she generates a descending sequence of integer values for the total number of people in the restaurant. By starting with 98, she takes into account some of the problem givens although she does not consider the fact that one table for four would be still available. In the next column, she calculates the division by three of the totals, in order to get one-third of the totals (the number of men). In another column, she calculates twice the previous results (the number of women). In the last three columns, another condition of the problem is addressed, namely, the distribution of customers by tables of three and of four. The successive multiples of four represent the varying number of people in tables of four, and as there was only one table with three people, the number three is repeated along another column. Then she adds the values in the previous two columns, which yields a column for the total of people. Comparing the two columns with the totals, she finds the number appearing in both columns, which gives her the number of people in the restaurant.

In solving the problem, the student uses the idea of proportion to "separate" the customers by gender, as mentioned in her answer: "Since the number of women is exactly twice the number of men, it can be concluded that the total of persons is represented as 3-thirds, being one third of men, and two thirds of women". She also uses the notion of multiples of four to define the number of people sitting at tables of four and a column with the fixed number three to act as a constant standing for the three people seated at one table.

It is apparent that using the spreadsheet pushed her to identify all the relevant variables and constants and encouraged the search for dependency relations. In addition, it led to a strategy that allowed addressing the two conditions involved in the problem separately and later making their connection by finding equal outcomes in the two independent tables created (Fig. 6.21).

Ana, in the same class, took a different approach from those of her classmates, as illustrated in Fig. 6.22.

Ana began by considering the condition that relates the number of men with the number of women, and after, in a third column, she obtained the total number of people by adding the values of the previous two variable-columns. The student concluded that the totals were multiples of three. Then she subtracted three from the total of persons to account for the fact that only one table had three individuals and divided the result by four (thus distributing the resulting numbers by tables of four).

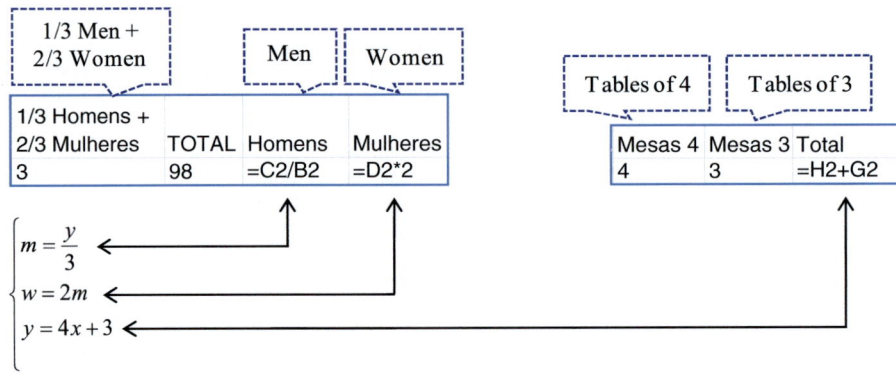

Fig. 6.21 Translation into algebraic language of Carolina's model on the spreadsheet

number of men	number of women	number of people without the 3	number of tables of 4	number of men	number of women	number of people without the 3	number of tables of 4
nº de homens	nº de mulheres	total	nº de pessoas sem as 3	nº de mesas de 4	nº de homens	nº de mulheres	total	nº de pessoas sem as 3	nº de mesas de 4
1	2	3	0	0	1	2	3	=F4-3	=H4/4
2	4	6	3	0,75	2	4	6	=F5-3	=H5/4
3	6	9	6	1,5	3	6	9	=F6-3	=H6/4
4	8	12	9	2,25	4	8	12	=F7-3	=H7/4
27	54	81	78	19,5	27	54	81	=F30-3	=H30/4
28	56	84	81	20,25	28	56	84	=F31-3	=H31/4
29	58	87	84	21	29	58	87	=F32-3	=H32/4
30	60	90	87	21,75	30	60	90	=F33-3	=H33/4
31	62	93	90	22,5	31	62	93	=F34-3	=H34/4
32	64	96	93	23,25	32	64	96	=F35-3	=H35/4
33	66	99	96	24	33	66	99	=F36-3	=H36/4

xls file   xls file (command "show formulas")

Fig. 6.22 Excerpt of Ana's solution in the spreadsheet (classroom student)

In cell H3, Ana entered the title "Number of tables of 4", and in the line below, she created the formula "=G4/4" and then dragged the fill handle of the cell. This way, she was finding the number of tables of four people that were in the restaurant. Finally, she just had to inspect the values in column H to look for integers and to find the highest number that would correspond to a total of people not above 100.

In her answer, Ana wrote: "The maximum number was 87 for this was before the 4 friends came, if I considered the number 99 and added the 4 friends I would get 103 in total but the capacity of the restaurant is 100 people, which means it is not the solution". Her ingenious thinking allowed her to get even more information than the previous solutions could achieve: the number of men, the number of women, the number of tables of four people and the total of people already in the restaurant.

6.3 Data Analysis

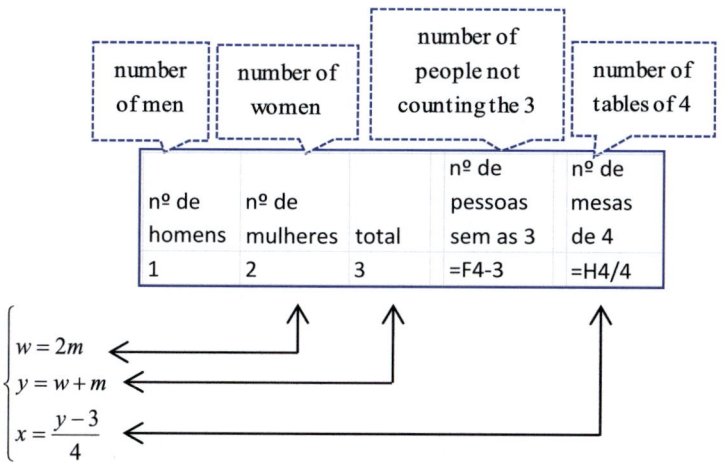

Fig. 6.23 Translation into algebraic language of Ana's model in the spreadsheet

In Fig. 6.23, we can see how Ana expressed the conditions in the spreadsheet and how they were chained in a particular arrangement without separating them, as was the case in the previous examples. We can also compare Ana's work in the spreadsheet with the corresponding conditions written in algebraic language.

The several solutions presented show the importance of the identification of all the variables in the problem and the conditions connecting them. The work with the spreadsheet enables students to deal with different conditions' formulation and to experimentally determine the solution of a system with several equations. In the four cases analysed, we find a clear image of how it is possible to generate different equations that may translate the given problem into algebra.

Our claim is that it demonstrates how the spreadsheet offers a rich environment to develop different and productive conceptual models in dealing with many interconnected variables involved in the conditions that frame the problem situation. Each of those conceptual models is reflected in a specific way when expressed through the computational tool. Devising different approaches and models with the use of the spreadsheet is an important step in the development of subsequent symbolic approaches and standard algebraic models. As different equations may represent the same problem, this means a promising path to engage students in realising how they can be transformed into others and thus uncovering instances of algebraic transformations in a set of conditions.

When trying to rank the solutions to this problem based on the criterion used before, we conclude that it is an insufficient criterion to distinguish them. In fact, all the examined solutions show that students made use of variable-columns, which indicates a more advanced appropriation of the tool and a more efficient use of the number variation feature offered by the use of the spreadsheet. Nevertheless, other possible criteria would be imaginable to undertake the task of creating a progression in types of solutions, which is foreseen as part of our future research development.

6.4 Discussion and Conclusion

The two problems that were proposed in the competition SUB14 are examples of the kind of global/metalevel activities considered by Kieran (2004), insofar as they involve functional reasoning and pattern-finding strategies. They both have in common the search for a maximum value, leading to some difficulties when a purely algebraic approach is envisioned. However, a spreadsheet provides alternative approaches to both problems that may make them clearer to students, facilitating their solution process and efficiently providing a path to the solution. We have examined the digital representations used by the students (in and out of school) as well as how they coped with them to get the solutions. On the other hand, we also recognised relevant aspects of algebraic thinking in students' activity and productions.

When analysing the various solutions made in the spreadsheet, we find a pattern of similarities, even with students acting in very different contexts (in the classroom or out of school within a web-based mathematical competition). All of them include a table with a certain "organisation" fostered by the affordances of the spreadsheet. In each of the solutions presented, the students identified the conditions that framed the problem situation and expressed them in the specific language of the spreadsheet. They recognised the relevant variables and through the definition of variable-columns or recipient cells expressed the relationships connecting several variables, some depending on others, in sometimes composite relations. Furthermore, it is possible to observe a correlation between each of these conditions expressed numerically in the spreadsheet and the algebraic conditions that make a system of equations with inequalities of integers also involved.

In addition, all students, whether in the classroom or participating in the online competition, include an explanation/justification of their procedures as an integral and fundamental part of their problem-solving and expressing, which crystallise students' conceptual models of the problem situations (Lesh & Doerr, 2003). These explanations/justifications concern mainly the assignment of variable-columns or recipient-columns, the establishment of relationships between columns by means of formulas and the inspection of the table values in search for the solution. In some cases, as in the scheme presented by one of the students, it becomes clear how the work in the spreadsheet connects and gives meaning to its translation into algebraic language, thus leading to expressing in algebraic formal mode.

We have been able to unfold students' problem-solving and expressing with the spreadsheet, sometimes combined with other commonly used digital tools, within a digital-mathematical expository discourse—a discourse where operations performed with the spreadsheet come to the fore but where, at the same time, the tool is the main medium for developing and expressing the mathematical (namely algebraic) thinking. We therefore consider these students as instances of humans-with-media, in the sense that their conceptual models are intrinsically shaped by the tool they chose to resort to but also in the sense that such models are portrayed (expressed) through a particular computational structure.

6.4 Discussion and Conclusion

For example, decisions such as the ways of defining variable-columns are intrinsically connected to the individual's conception of the way in which variables depend on other variables and to the interpretation leading to the choice of independent and dependent variables. Therefore, the different spreadsheet organisations appearing in a particular problem are a consequence, among other things, of conceptual choices and decisions and constitute a powerful mirror into students' productive ways of conceiving the problem situation. Such decisions have to be seen as a permanent interaction between the subject and the tool that makes it impossible to separate the two. Evidence of such interdependence is quite clear when we observe students in their face-to-face ongoing problem-solving activity and likewise are well in sight on the answers given online by the competition's participants, as long as we look at them as solving and expressing products.

In the first problem, in all of the students' models, four columns corresponded to the four sons' number of bars, and the column attributed to the second son's number of bars was intended for the introduction of initial values (the inputs), representing a variable-column that works as an independent variable. The remaining columns were constructed through relations of dependence, in some cases by using formulas and in others just by manually entering values after calculations were performed externally. Either using trial and refinement or creating a continuous sequence of integers for the second son's number of bars, the students carried out the inspection of a control column assigned to the total number of delivered bars, which meant checking one last condition in the problem (1000 is the total number of bars to be distributed among the four).

In the second problem, we found a greater variety of approaches. On the one hand, we have solutions in which students turn to a strategy where they consider all the conditions of the problem globally (one single system of equations), and on the other hand, we have other students using the strategy of separating two sets of conditions (two subsystems of equations) and then comparing the numerical values for the same quantity, in each set of columns, to find the common numerical result.

In general, we found that the spreadsheet helped students to deal with intricate algebraic problems, beyond their formal and technical mathematical knowledge, by means of sophisticated numerical approaches based on translating relations between variables into numerical sequences, namely, through the use of variable-columns. We claim that algebraic thinking was fostered by the affordances of the spreadsheet in structuring and giving a medium to express the rules inside the problems. This result resonates with other studies such as Ainley et al. (2004), but it also underlines the structure of students' algebraic thinking expressed in a particular representation system. It provides a clear indicator of how students interpreted the problems in light of their mathematical knowledge and their knowledge of the tool.

The analysis allows us to make inferences about what is gained in using the spreadsheet to solve algebraic problems and helps us to understand an enormous plasticity in the relationship between the symbolic language of the spreadsheet and the symbolic language of algebra. The use of Excel can be seen as a means to fill the gap between the algebraic thinking and the ability to use algebraic notation to express such thinking. The kind of algebraic thinking that emerges from the use of the spreadsheet is the kind that belongs to global algebraic activities (Kieran, 2004).

Finally, some of the features and roles of the spreadsheet in algebraic problem-solving should be reiterated

1. The spreadsheet environment proves to be a useful environment to anticipate complex algebraic problems; our study shows how the spreadsheet allowed 7th and 8th graders to solve two problems before they had the formal knowledge of specific algebraic topics. On the other hand, it anticipated forms of algebraic reasoning involved in the problems that were elicited by the representation systems embedded in the spreadsheet.
2. The spreadsheet and the user are involved in a co-action process in constructing relations between the several variables involved and in expressing conditions and restrictions appropriately. These were mainly crystallised in the language of the spreadsheet but inducing a quasi-symbolic algebraic language.
3. The spreadsheet offered a digital approach, in the sense that it means a numerical character and an experimental approach to an algebraic problem, especially visible in students' ways of representing the problem through numerical variable-columns.

The use of the spreadsheet in problem-solving can offer a bridge between arithmetic and algebra as it offers an environment that the two fields naturally cohabit. It strengthens the understanding of variation and variable, including ways of thinking about the relation between a set of inputs and a set of outputs and of formulating relations that are a composition of other relations.

In the work of students in the classroom, it was quite apparent that most of the problem-solving was the result of a permanent co-action with the spreadsheet. However, there are situations where the observed co-action is somehow limited, as it was apparent in the first problem, in which some of the students did not use formulas and most of the calculations were done outside the spreadsheet by the students themselves.

In general, the various forms of co-action revealed by students' solutions can be identified with specific types of co-action that are typical of different stages of computational development in the light of the proposal made by Moreno-Armella et al. (2008). Thus, we would put the types of co-action shown as typical of the third and fourth stages.

After reviewing the two problems and taking into account a number of other studies (e.g. Ainley et al., 2004; Carreira, 1992; Dettori et al., 2001; Haspekian, 2005; Nobre, Amado, & Carreira, 2012; Rojano, 2002; Tabach et al., 2008), we have an empirical basis to conclude that the affordances of the spreadsheet can be mathematised, as suggested by Hegedus (2013), and we can also add that such harnessing is actually done by young students (as we have shown with 7th and 8th graders) that deliberately make the choice of using this digital tool, both inside and outside the mathematics classroom.

After addressing the points we set out to investigate, we are ready to raise new questions and make suggestions that may direct future research. One point that we could not address, and one that would be worthy of deeper attention, is to look for possible connections between students' mathematical knowledge, particularly in the field of algebra, and the degree of complexity of their use of the spreadsheet and the degree of generality of their solutions (something close to the idea of formal models). Another possibly useful effort would be the development of a classification of different types of word problems according to the opportunities they offer in terms of the development of algebraic concepts and symbolic language and simultaneously according to its suitability to some affordances of the spreadsheet, thereby providing a resource for teachers and teacher educators.

Finally, we have proposed a first attempt at possible ways of ranking different spreadsheet-based solutions. This preliminary suggestion was able to indicate a hierarchy in a set of solutions to the first problem ranging from using recipient cells to using variable-columns. Other criteria are obviously possible and may be combined with the former in a way that may extend our knowledge on the ways young students use the spreadsheet to solve problems on quantity variation. This is a further step that we are aiming to take in our future research.

References

Ainley, J., Bills, L., & Wilson, K. (2004). Constructing meanings and utilities within algebraic tasks. In M. J. Høines & A. B. Fuglestad (Eds.), *Proceedings of the 28th PME Conference* (Vol. 2, pp. 1–8). Bergen, Norway: PME.

Carreira, S. (1992). *A aprendizagem da Trigonometria num contexto de aplicações e modelação com recurso à folha de cálculo* (Master Dissertation). Lisbon, Portugal: APM.

Dettori, G., Garuti, R., & Lemut, E. (2001). From arithmetic to algebraic thinking by using a spreadsheet. In R. Sutherland, T. Rojano, A. Bell, & R. Linz (Eds.), *Perspectives on school algebra* (pp. 191–208). Dordrecht, The Netherlands: Kluwer.

Haspekian, M. (2003). Instrumental approach to understand the problems of the spreadsheet integration. In T. Triandafillidis & K. Hatzikiriakou (Eds.), *Proceedings of the 6th International Conference Technology in Mathematics Teaching* (pp. 118–124). Athens, Greece: New Technologies Publications.

Haspekian, M. (2005). An "instrumental approach" to study the integration of a computer tool into mathematics teaching: The case of spreadsheets. *International Journal of Computers for Mathematical Learning, 10*(2), 109–141.

Hegedus, S. (2013). Young children investigating advanced mathematical concepts with haptic technologies: Future design perspectives. *The Mathematics Enthusiast, 10*(1–2), 87–108.

Kieran, C. (1996). The changing face of school algebra. In C. Alsina, J. M. Alvares, B. Hodgson, C. Laborde, & A. Pérez (Eds.), *8th International Congress on Mathematical Education: Selected Lectures* (pp. 271–290). Seville, Spain: SAEM Thales.

Kieran, C. (2004). The core of algebra: Reflections on its main activities. In K. Stacey, H. Chic, & M. Kendal (Eds.), *The future of teaching and learning of algebra: The 12th ICMI Study* (pp. 21–34). Dordrecht, The Netherlands: Kluwer.

Kieran, C. (2007). Developing algebraic reasoning: The role of sequenced tasks and teacher questions from the primary to the early secondary school levels. *Quadrante, 16*(1), 5–26.

Lesh, R., Behr, M., & Post, T. (1987). Rational number relations and proportions. In C. Janvier (Ed.), *Problems of representation in the teaching and learning of mathematics* (pp. 41–58). Hillsdale, NJ: Erlbaum Associates.

Lesh, R., & Doerr, H. (Eds.). (2003). *Beyond constructivism: Models and modeling perspectives on mathematics problem solving, learning, and teaching*. Mahwah, NJ: Erlbaum Associates.

Lins, R., & Kaput, J. (2004). The early development of algebraic reasoning: The current state of the field. In K. Stacey, H. Chick, & M. Kendal (Eds.), *The future of teaching and learning of algebra: The 12th ICMI Study* (pp. 47–70). Dordrecht, The Netherlands: Kluwer.

Moreno-Armella, L., Hegedus, S., & Kaput, J. (2008). From static to dynamic mathematics: Historical and representational perspectives. *Educational Studies in Mathematics, 68*, 99–111.

Moreno-Armella, L., & Hegedus, S. (2009). Co-action with digital technologies. *ZDM: The International Journal on Mathematics Education, 41*, 505–519.

Nobre, S., Amado, N., & Carreira, S. (2012). Solving a contextual problem with the spreadsheet as an environment for algebraic thinking development. *Teaching Mathematics and its Applications, 31*(1), 11–19.

Ozgun-Koca, S. A. (2000). *Using spreadsheets in mathematics education*. Office of Educational Research and Improvement (ERIC Digest). Retrieved from http://www.ericdigests.org/2003-1/math.htm.

Radford, L. (2000). Signs and meanings in students' emergent algebraic thinking: A semiotic analysis. *Educational Studies in Mathematics, 42*(3), 237–268.

Rojano, T. (2002). Mathematics learning in the junior secondary school: Students' access to significant mathematical ideas. In L. English, M. B. Bussi, G. A. Jones, R. A. Lesh, & D. Tirosh (Eds.), *Handbook of international research in mathematics education* (Vol. 1, pp. 143–161). Mahwah, NJ: Erlbaum Associates.

Tabach, M., Hershkowitz, R., & Arcavi, A. (2008). Learning beginning algebra with spreadsheets in a computer intensive environment. *The Journal of Mathematical Behavior, 27*(1), 48–63.

Weigand, H., & Weller, H. (2001). Changes in working styles in a computer algebra environment: The case of functions. *International Journal of Computers in Mathematical Learning, 6*(1), 87–111.

Wilson, K. (2007). Naming a column on a spreadsheet. *Research in Mathematics Education, 8*, 117–132.

Zazkis, R., & Liljedahl, P. (2002). Generalization of patterns: The tension between algebraic thinking and algebraic notation. *Educational Studies in Mathematics, 49*, 379–402.

Zazkis, R., & Liljedahl, P. (2004). Understanding primes: The role of representation. *Journal for Research in Mathematics Education, 35*(3), 164–186.

Chapter 7
Digitally Expressing Co-variation in a Motion Problem

Abstract Co-variational reasoning has received particular attention from researchers and mathematics educators because it is considered of paramount importance for the understanding of concepts such as variable, function, rate of change, derivative, etc. Some of the critical issues that have been identified in several studies consist of the difficulty in interpreting the simultaneous variation of two quantities, particularly in overcoming coordination problems of two variables changing in tandem. A relevant question in the study of co-variational reasoning concerns representing the joint variation of quantities and performing translations between different representations. Problems of motion involving variation over time are strongly linked to the concept of co-variation and require the ability to translate a dynamic situation by means of mostly static representations. Those problems require the construction of a conceptual model that, in some way, visually contains dynamism. In taking solving and expressing as a unit of analysis and focusing on the ways in which commonly available digital technologies are used by youngsters as tools in problem-solving, we analyse the approaches used by the participants in SUB14 to a motion problem. Some surprising results of the content analysis of over 200 answers indicate that the textual/descriptive form of presenting a model of the situation had a clear dominance. The use of tabular representations along with pictorial/figurative content was also present in a high percentage of solutions. Furthermore the use of digital media was decisive in producing visuality, i.e. ways of depicting the displacement with time (quasi-dynamic representations).

Keywords Co-variation • Co-variational reasoning • Motion problem • Conceptual models • Representation modes • Visuality • Pictorial/figurative representations • Common usage digital tools

7.1 Main Theoretical Ideas

The mathematical competitions SUB12 and SUB14 each offer a variety of challenging problems to the young competitors along their two phases (see Chap. 1). The choice of the problems posed within the competitions is not subordinate to the official mathematics curricula nor is it constrained by the national curricular targets or formal assessment. An important criterion behind the choice of the problems concerns the *moderate challenge* that is deliberately intended; another is envisaging ways in which students may tackle the problems apart from conventional and formal mathematical knowledge learned in school. This means that some problems may belong to areas of mathematical knowledge that participants are likely to encounter only much later in their school attendance—as part of specific content matter or curricular topics (like the case of combinatorial, only taught in secondary school)—but considered to be within the reach of the youngsters through informal processes emerging from their mathematical thinking and conceptual models development upon contextual problem situations.

In this chapter, we focus on a motion problem, involving the co-variation of displacement and time in a relative motion situation, posed during the online phase of the SUB14. The decision to focus on a motion problem, among many other problems proposed along the SUB14 editions, has to do with the fact that a problem involving motion—while relating the variables space and time—requires some kind of understanding of the dynamic nature of the situation and finding suitable models for their representation. We therefore examine the problem-solving and expressing of the young participants when facing a motion problem, especially taking into account that most participants employ some sort of digital medium to express their thinking.

The data refers to all the digital solutions submitted by the 8th graders participating in SUB14 in the 2011–2012 edition of the competition. Besides the fact that the whole set of answers (covering both 7th and 8th graders) was found too extensive to allow a feasible qualitative data analysis, it was decided to focus only on 8th graders because those students, compared to 7th graders, are expected to have gained more training and ease with algebraic methods in their school mathematics (namely, on establishing and solving equations). This suggested the possibility of receiving more answers using algebraic methods, namely, equation-based solutions, from 8th graders than from 7th graders. That was seen as a relevant factor when looking at the different strategies and ways of solving the problem presented by the participants. In fact, 7th graders may not choose to deal with equations in solving the problem because of a limited knowledge on how to use equations, and this could become the main reason for devising another solution process, whereas 8th graders are expected to be more apt to do equations and thus would have more mathematical tools to consider when deciding on their approach to the problem.

Our aim in this chapter is to describe and characterise youngsters' approaches to a problem situation where there are two bodies starting from opposite ends and moving towards each other, with different speeds, along with particular conditions that allow determining the time at which the bodies meet. We also want to examine students' modelling of a motion situation in relation to the technological tools they have used to express their understanding of the problem.

The data analysis was planned to be developed in two stages: first, the problem was solved by a group of experts (mathematicians) and their solutions were categorised according to the strategies developed and the conceptual models imbedded in their approaches to the problem, and then the a priori categories were used to sort and classify the 8th graders' solutions. The interpretation of the participants' categorised data is expected to generate insights into students' prevalent conceptual models. Students' models and strategies in solving the motion problem were also investigated in connection to the technological tools they employed in their solving and expressing of co-variation.

We now expand on the idea of modelling motion and, in particular, the notion of co-variation. We then say more about visualisation when tackling motion problems.

7.1.1 Co-variation and Modelling Motion

Creating models of continuously changing events or understanding dynamic functional situations are examples of capabilities that may be subsumed under the notion of co-variational reasoning (Zeytun, Çetinkaya, & Erbaş, 2010). This type of mathematical reasoning is mainly concerned with developing good images that involve coordinating simultaneous changes of more than one variable. According to studies conducted on understanding change, rate of change, functional variation, derivative and other related concepts, one of the key difficulties involved in co-variational reasoning is to grasp the simultaneous change of different quantities and the effects of changing one quantity on another quantity. Such findings have supported pedagogical recommendations to anticipate the study of functions to earlier ages and to address the teaching of functions and change from the point of view of co-variational thinking (Carraher, Martinez, & Schliemann, 2007).

An important point about the difficulties associated with co-variational reasoning in mathematics teaching and learning is that teachers tend to consider this kind of reasoning and the modelling of dynamic situations as very complex and not easily accessible to their students. For example, the study by Zeytun et al. (2010) showed that teachers are more accustomed to thinking of functions as a correspondence than as a co-variation; also their predictions about students' failure in co-variation tasks matched the high level of difficulty they themselves experienced on those tasks.

While acknowledging the complexity of co-variation, other researchers have also argued that developing students' conceptual models to interpret, represent and model dynamic situations is an important pathway to overcome coordination problems of two variables changing in tandem (Carlson, 2002; Carlson, Jacobs, Coe, Larsen, & Hsu, 2002; Carlson, Larsen, & Lesh, 2003). Some specific tasks have become well known for their use in investigations that focus on issues of coordination of the variables changing with respect to each other and coordination of continuous changing rate: such is the case of problems on graphing the water level changing while imagining the filling of a bottle with a particular shape. Carlson et al. (2003) have adapted these prototypical tasks to create model-eliciting activities designed to promote conceptual development of co-variational reasoning and concluded that the solutions given by pre-service elementary teachers revealed they were able to develop and refine a general model for analysing the dynamic situation involved.

Rather than concentrating on plotting point after point to construct the graph, they used the rate of change as the way to make sense of the dynamic situation, drawing on informal thinking in exploring the problem.

One central aspect of co-variational reasoning, according to Carlson et al. (2002), is its close connection to mental imagery in the sense that co-variation requires constructing images of dynamic, changing, moving phenomena, grounded on physical actions and bodily movements, like higher or lower, rising or declining, faster or slower, closer or further away, etc., which are carriers of metaphorical meaning about the changing of variables in tandem with each other. Moreover, as the results of their study showed, individuals often try to "simulate" the dynamic situation, enacting the problem through the use of objects, body and gestures. The following were findings revealing how such materialisation of the dynamic situation became apparent in participants' answers to the problem of describing the speed of the top of a ladder sliding down a wall as the bottom is pulled away at a constant rate:

> When one of these students (Student B) was prompted to explain his correct response, he performed a physical enactment of the situation, using a pencil and book on a table. As he successively pulled the bottom of the pencil away from the book by uniform amounts, he explained, "As I pull the bottom out, the amount by which the top drops gets bigger as it gets closer to the table" (…). Student A provided a similar response, except that her enactment involved using her hand and a book to model the situation. She began by pressing her flat hand against the book and successively moved the bottom of her hand away while watching the amount by which the top of her hand dropped down. (Carlson et al., 2002, pp. 371–372)

Therefore, co-variational reasoning appears to be strongly related to developing real and kinaesthetic images of the dynamic situations being modelled. The use of physical enactment was pointed out by Carlson (2002) and Carlson et al. (2002) as a powerful representation tool to support co-variational thinking. Likewise, Matos and Carreira (1997) have studied students' making sense of co-variation on a modelling problem involving the calibration of a double-coned-shaped log glass. The study suggested that students' thinking and meaning-making concerning the simultaneous change of variables such as level, volume and time were permeated by multiple instances of metaphorical and analogical reasoning.

Although the co-variation of two quantities does not always require the notion of time, the metaphor of the exact time for the location of a moving point has often been helpful to discuss and analyse co-varying quantities. In summary, co-variational reasoning holds a consistent connection with the creation of dynamic mental images, metaphorical reasoning, physical enactment and bodily referents.

The mathematics of change, namely, concepts involving variables and functions, is also characterised by the many interrelated representation systems that allow capturing different meanings and notions depicting variation, namely, tables, graphs and plots. The coordination as well as the conversion between these different representational systems is at the heart of functional and co-variational reasoning.

Passaro (2009) developed a study with students aged 13–14 where she proposed a problem on studying the co-variation between "distance travelled" and "displacement" during a certain walk to be defined on a map of a neighbourhood, assuming

7.1 Main Theoretical Ideas

that the independent variable should be the distance covered. In solving the problem, the students were asked to give a written description of the co-variation and to produce a visual representation of the phenomenon described. This case is of importance for the present study because it involves a motion problem and also because it gives particular attention to visualisation and forms of expressing in students' answers, to which the researcher refers as students' spontaneous representations (Passaro, 2009). The results were prominent in revealing a large variety of visual forms for representing the situation:

> First of all it is interesting to observe the variety of representations presented. In fact, students had to be creative to be able to visually represent the phenomenon of covariation. (Passaro, 2009, p. 68)

Moreover, in that study, the several collected representations were distinguished and grouped according to their degree of abstraction. The three levels that came out of the analysis evolve from a level that tries to depict the situation, including irrelevant details of the walk and limited information on the variables, to a level of more schematised representations, including sound information concerning the two variables.

Finally, the study brought in striking results on the effectiveness of the two representation modes, i.e. verbal description and visual registers, by showing that visual representations were an obstacle to students' expressing the distinction between the two variables and even more ineffective in considering the simultaneous change of the variables.

> As to the written descriptions, it appears that more than half of the students seem to identify and distinguish two variables (27 students out of 50), but less than a quarter (7 out of 50 students) establishes a dependency relationship between these two variables. In what concerns the visual representations, they show a poorer perception of these significant elements as only 12 students visually represent the two quantities considered and 2 are able to represent dependency. (Passaro, 2009, p. 71)

Even though the above problem involves a dynamic situation, i.e. motion, it is noteworthy that the variable time is omitted and it only implicitly has an effect on the ways in which the variables change in relation to each other. In this respect, the study from Matos and Carreira (1997) has revealed that using composite functions provided a useful tool to develop clearer images of co-variation. In the example of the co-variation between displacement and distance in the walk situation, if one admits that distance increases with time at a constant rate, $s(t)$, then picturing the change of displacement with time, $r(t)$, will help in creating a reasonable image of the change of displacement with distance, $r(s)$.

Existing research thus points to several critical aspects of co-variational reasoning. The greatest difficulties are centred on the need to identify the relevant variables, the importance of coordination and the requirement to create images of the effect of changing one variable in relation to the other. The way motion is represented seems to be decisive and closely linked to the construction of visual images, often supported by gestures and physical enactment.

What is not clear, however, is the role of different representational means and tools in helping students to express co-variational reasoning. In particular, further thinking on the visualisation component involved in the mathematisation of motion and dynamic functional situations is apparently needed.

7.1.2 Visualisation in Motion Problems

Research on co-variational thinking has been recognising the centrality of the representational issues, particularly in studies concerning solving algebraic problems and also the mathematical modelling of dynamic situations by functions and equations. Our own research aims also converge to the inspection of representational elements in youngsters' solving of a motion problem, by considering their strategies and the nature of the representations they create in their processes of achieving a solution.

As pointed out by Izsák and Findell (2005), a problem involving relative motion, where data concerning speed and relative positions of two moving bodies are given and the time taken for the two bodies to meet is wanted, can be seen as an algebraic problem where students need to shift attention from a varying quantity in a situation to a specific value of that quantity (traditionally seen as finding the unknown through formulating an equation and solving it). One of the problems discussed in their study refers to the total number of bags of popcorn sold over time, during a day. An important shift of attention was required between considering the total bags sold at a particular time (the cumulative quantity) and the number of bags sold each hour (the rate quantity). The representational features that were considered to support adaptive interpretations were rooted on arithmetic representations such as using the number line and horizontal arrows (vectors) to represent changes per hour, which in turn could suggest $x-y$ graphs where the increments on the y-axis could be pictured in the form of vertical distances (vertical arrows or vectors). Adaptive interpretations are examples of representational features that seek to connect students' experiences in algebra with earlier experiences in arithmetic in problems involving co-varying quantities (Izsák & Findell, 2005).

The idea of developing forms of representing co-variation rooted in more arithmetical ones, as in the case of a horizontal timeline, suggests a form of visualisation and of coordination that is of particular relevance to our analysis of students' spontaneous representations when they employ commonly used technological tools. In students' digital solutions to motion and dynamic problems included in our empirical data, the use of arrows, lines, dots, captions, labels, etc. has been seen to be one of the usual ways that students find to digitally express movement and related mathematical ideas. If understood as a certain facet of the digital world and the digital gaming that many youngsters intensively manipulate, this type of adaptive interpretation may reflect their own ways of linking co-variational thinking to earlier experiences.

7.1 Main Theoretical Ideas

Monaghan and Clement (2000) have also addressed students' gains in producing mental imagery during the solution of relative motion problems when using a computer animation tool. When comparing a group who had an animation treatment with a group who had a numerical with static graphic treatment, the researchers found many imagery indicators in the first group, which included self-projection, depictive hand motions and verbal descriptions of relative speed and position. Furthermore, the data on the numerical treatment group showed "evidence for the construction and use of faulty mechanical algorithms, a striking contrast to the mental imagery evidenced by animation students" (Monaghan & Clement, 2000, p. 319).

As a conclusion of the study, the focus on numerical data tend to prompt mechanical algorithmic approaches in solving relative motion problems, whereas the use of visual animations fosters visualisation even when motion problems are solved without the use of the computer animation. This is significant for our research purposes since we are looking at students' expression of relative motion through the use of technological tools, although not with the support of animation tools but rather taking advantage of their own daily technological resources.

Of no less importance is the perspective brought to the forefront by Sherin (2000) who set forth the study of the representations of motion that young people invent, taking into account the whole representational background they get since childhood. In the view of Sherin, we must look at what students invent, create and produce, in light of their representational experience. Indeed, we support the claim that today's young generation has acquired a representational experience that is not inconsiderable when it comes to scrutinising the representations they create on their personal computers to express motion.

> From the time they are infants young children are exposed to a wide variety of representations. Thus, whenever students learn to use a new representation, it is learned against this background of experience. What is missing in our accounts, I believe, is an attempt to see the learning of representations in this broader light. I refer to the new approach as "genetic," because it attempts to understand representational learning within the broader context of the genesis of representational competence. (Sherin, 2000, p. 400)

The researcher's priority of looking at children's capabilities rather than at their difficulties or deficits is also of interest. It is echoed in his remark on the tendency to underestimate the drawings and the pictures used by children to represent motion, often regarded as not useful and contrary to standard scientific representations:

> But I believe that such a negative attitude toward drawing is not justified, and that there is much that the ability to draw can contribute to the use of the standard scientific representations. (Sherin, 2000, p. 414)

Sherin's genetic analysis of students' drawings in representing motion and relative motion resonates with much of the evidence we have collected from students' work on motion problems making use of ordinary technology, such as Paint, the drawing tools in Word, or PowerPoint items to create diagrams. Like in the case of the drawings, temporal sequences and graph-like representations found in his study (Sherin, 2000), the ways in which the participants in SUB12 and SUB14 combine dots, lines, arrows, pictures, images, labels and colours in their digital productions, are indicators that student-invented representations can be important factors of meaning-making in motion problems, supporting visualisation and coordination in co-variational thinking.

As Sherin (2000) has claimed, we want to consider students as designers of representations and to look thoroughly at how such creative representations are integral parts of the skeletons for their conceptual models of relative motion:

> (…) inventing representations can do more than help students to understand existing representations; the ability to invent representations can be a useful skill in its own right. This is particularly true given the increased prevalence of technological tools. The wide availability of computers has allowed an expansion in the quantity and variety of representations that people are faced with on a daily basis, and in the opportunities for individuals to create their own representations. We want students to have the skills to negotiate this representational fray, and themselves to become disciplined designers of representations. (Sherin, 2000, p. 438)

7.2 Context and Method

The aim of the analysis presented in this chapter is to understand how the conceptual models and types of mathematical representations that students used to solve and express a motion problem are related. It is important to emphasise that the documentary data analysed (the solutions sent by e-mail by the participants) present the ways in which students express their processes to reach the solution of the problem as this is a requirement imposed by the competition rules for an answer to be considered correct.

Surely, the modes of explanation, i.e. the expository discourse of young participants vary in aspects such as the degree of development of the explanation submitted, the more or less formal character of mathematisation (from the point of view of mathematical language), clarity, organisation, etc. However, the answers under analysis satisfy the requirement of providing an account of the process carried out to achieve the solution. It should also be noted that participants are never asked to find more than one approach to the same problem and, as such, it is extremely rare to obtain answers that include more than one form of solving the problem. Finally, it should be noted that participants may have reached the correct and complete solution after more than one submission as they always receive feedback by e-mail and have the opportunity to refine and resubmit their answers.

The research purpose is to know how students deal with a problem that concerns space and time co-variation, paying particular attention to their ways of modelling the situation and to the features of their mathematical representations within the media they choose to express their thinking.

The motion problem was launched early in the SUB14 competition, being the first of the problems proposed in 2011–2012. Given the large volume of answers received from all participants involved in the competition, which makes it excessive for a comprehensive qualitative analysis, the choice was to focus the analysis on the answers received from the participants attending 8th grade. This decision limited the dataset analysed to a total of 254 answers to the problem. Of these answers, 220 were classified as correct, all containing a full explanation of the process used to obtain the solution. The remaining ones were incorrect or did not include any

7.2 Context and Method

Alexander and Bernard live at a distance of 22 km from one another and want to meet but have only one way to make the journey... it is by walking! On a holiday morning they decided to walk towards one another to get together. Alexander left his home at 8 a.m. and went walking at a speed of 4 km per hour. Bernard left his home an hour later and walked at a speed of 5 km per hour. Neither of the two friends took his watch but we can know the time they met each other. What time was it?

Do not forget to explain your problem solving process!

Fig. 7.1 Problem #1 from the SUB14 (edition 2011–2012)

grounds on how the solution was achieved. Although this was the first problem of the season, many of the participants were already used to sending attached files with their presentation of the solution (from previous years). The different formats that were obtained in terms of the digital tools used include, in order of prevalence, text editor, presentation editor, image editor, spreadsheet and GeoGebra. There were also several PDF files sent, most of which appear to result from an original file in Word or another word processor.

The problem mentions two individuals walking towards each other with different speeds and different departure times that will meet each other eventually. The problem statement is concise and includes an illustrative picture, following the usual format of the problems presented throughout the competition (Fig. 7.1).

In order to create categories to describe students' ways of thinking and representational practices, we adopted a strategy also developed in other research studies (e.g. Santos-Trigo, 1996, 2004). We started by looking for different solutions to the problem to steer the identification of strategies and models and suggest clues to develop the categorisation of students' answers. Our approach was to ask five experts to solve the problem. These five experts were chosen among university mathematics teachers from two Portuguese universities under the criteria of being openly interested in challenging mathematical problems and enthusiastic about the value of problem-solving in mathematics education.

The task proposed to each of the five experts had the following two requirements:

1. Solve the problem in the way that is most obvious and immediate to you (using paper and pencil or using any other tool, if wished).
2. After having the most direct way of solving the problem, think of other possible ways and present them (possibly one or two).

The five experts sent their answers by e-mail or handed them out written on paper. None of them used any technological tool to solve or express the problem, except the e-mail device itself when the answers were not handed out. However, one of them mentioned in the e-mail that it would be possible to think of a way of producing an animated solution for the problem by using the software Mathematica. They were all asked to deliver the solutions within 1 week. According to the order of reception of their solutions, the experts were assigned the letters A, B, C, D and E, for anonymous referencing.

7.3 Data Analysis

In what follows, first an analysis of how the problem was solved by a group of experts (mathematicians) is presented. Next, a categorisation is developed of the solution strategies used by the experts. Finally, the categorisation is used to sort and classify the 8th graders' solutions. In examining the 8th graders' solutions, our focus is on identifying the students' prevalent conceptual models and on their models and strategies with the technological tools they employed in their solving and expressing of co-variation.

7.3.1 The Experts' Solutions to the Problem

The five experts delivered, on the whole, 16 solutions: each one has produced three solutions, except for expert D who has presented four solutions. Thus, we have the following solutions: A1, A2 and A3 from expert A; B1, B2 and B3 from expert B; C1, C2 and C3 from expert C; D1, D2, D3 and D4 from expert D; and E1, E2 and E3 from expert E. Of the whole set, the solutions A1, B1, C1, D1 and E1 are those which the experts considered to be the most immediate for them. We now start by presenting a summary of the overall characteristics of the 16 solutions received from the experts.

Except in the case of expert C, who did not provide any type of drawing or diagram, all the others' first solutions begin to show a diagram consisting of a line segment limited by points A and B (all admitted that the trajectory of the friends Alexander and Bernard would be straight; only one expert made this assumption explicitly). Two of those diagrams show two arrows with opposite directions (vectors) originating from A and B and the respective speed values of 4 and 5 km/h.

In all the more immediate solutions (five solutions), the experts used an algebraic approach, establishing a linear equation, considering t as the unknown and getting the solution by standard algebraic methods. In all these algebraic solutions, the experts identified the variables space and time, and their equations reflected the direct proportionality between the distance travelled and the time elapsed. All solutions took as the central idea the fact that the sum of the distances travelled by Alexander and Bernard was equal to 22 km.

Only two of the experts (A and D) showed another type of approach beyond the algebraic algorithmic method to solve the problem. Expert A presented the solution A3 that combines a tabular representation and a schematic representation.

7.3 Data Analysis

Expert D presented the solution D2 based on a tabular representation; expert D further proposed the solution D3 which concentrates on an algebraic representation but is complemented by a graphical representation—this solution was the only proposal that included an $x-y$ graph to express the variation of space with time; and finally expert D offered the solution D4 that solely uses verbal statements to express the reasoning developed in finding the answer to the problem.

From the set of solutions presented, we identify two major conceptual models or ways of thinking about the problem situation (Santos-Trigo, 1996). These two models have in their basis a fundamental distinction: (1) to imagine the journey undertaken by the two friends as being already complete (i.e. they have already met at a certain point and time) and seek to obtain the elapsed time from the start until the moment they met, which means looking to the past, or (2) to imagine the journey undertaken by the two friends as it were happening chronologically (i.e. sequentially reconstruct the time and position of each of the two friends from the start) and seek to find the time and the position in which they will eventually meet, which means looking to the future. One expert refers to the second way to solve the problem as being "more constructive", and this reflects the fact of conceptually trying to recreate the journey undertaken by the two friends walking, step by step, rather than imagining it as being already completed or accomplished.

Thus, we propose two major categories of conceptual models that underpin the process of solving the problem:

Model 1: The completed journey (looking at the past)
Model 2: The developing journey (looking at the future)

With respect to model 1, of the completed journey, which predominated in the experts' solutions and was the basis of all the solutions that seemed more straightforward or immediate to them (i.e. A1, B1, C1, D1 and E1), their expression of the model occurs primarily through algebraic representations. It should be noted however that the algebraic representations, usually complemented by brief and concise textual representations, show variations in the ways of thinking about the situation.

Thus, we present a general outline of the various algebraic representations used in the expression of model 1, totalling six forms of algebraic representation. Although it is obvious that the equations used are equivalent to each other from a purely algebraic point of view, it is possible to observe that they express two distinct approaches. In one approach, it is quite clear the separation between two parts of the journey—the part where only Alexander walked (4 km walked and 1 h elapsed) and the part where the two friends walked together (after the first hour elapsed). In another approach, the separation between the two parts of the journey is not explicit, but instead it is a relationship established between the time taken by Alexander and the time taken by Bernard until they met (1 h of difference between the time they spent) (Table 7.1).

Model 1 also appeared in a solution of expert D (solution D4), which is identical to the solution in the second row of Table 7.1, but expressed through textual representations in everyday language, in a textual explanation of the mathematical reasoning, as follows: "At 9.00 Alexander has already walked 4 km and Bernard starts

Table 7.1 Summary of approaches and algebraic representations used with model 1

Model 1: the completed journey (looking at the past)		
Type of approach	Experts' solutions	Algebraic representation
Journey divided in two phases	A1	$4+(4+5)t=22$
	E1	
		Combined speed after the first hour
	B1	$4t+5t=18$
	C1	
	D4	
		Distance walked by the two friends after the first hour
		(D4: Informal algebra) "At 9.00 Alexander had already walked 4 km and Bernard was starting. The distance between them was then 18 km and the speed of approaching each other was 9 km/h (4 km/h + 5 km/h). Therefore, after 2 h, they have met each other, which means they met at 11.00"
	E2	$4+4t+5t=22$
		Distance walked by Alexander after 4 km made / Distance walked by Bernard from the moment he starts
	B3	$4+4t=22-5t$
	D3	
	E3	
		Distance walked by Alexander after 4 km walked / Distance walked by Bernard from the moment he starts
Relating the times taken by each of the two friends	D1	$4(t+1)+5t=22$
	B2	
	C3	
		Alexander walks one hour longer than Bernard
	A2	$4t+5(t-1)=22$
	C2	
		Bernard walks one hour less than Alexander

7.3 Data Analysis

Table 7.2 Summary of approaches and representations used with model 2

Model 2: the developing journey (looking at the future)			
Type of approach	Experts' solutions	Tabular and diagrammatic representations	
Numerical and geometrical	A3	1st hour	4 km
		2nd hour	4+9=13 km
		3rd hour	4+9+9=22 km
		$\underbrace{\text{4 km} \mid \text{4 km} \mid \text{4 km} \mid \text{5 km} \mid \text{5 km}}_{\text{22 km}}$ 1st hour, 2nd hour, 3rd hour, 3rd hour, 2nd hour	
Numerical	D2	Time (hours)	s_A | s_B | $s_A + s_B$
		8.00	0 | 0 | 0
		9.00	4 | 0 | 4
		10.00	8 | 5 | 13
		11.00	12 | 10 | 22

walking. The distance between them is then 18 km and the speed of approaching each other is 9 km/h (4 km/h+5 km/h). Therefore, after 2 h, they have met each other, which means they met at 11.00".

Regarding model 2, the developing journey, only the experts A and D have conveyed it, and it just appeared in some of their alternative solutions, as in both cases this model was not the basis of their most immediate and direct approaches for solving the problem. This model appears in the solutions A3 and D2 of experts A and D, respectively (Table 7.2).

Table 7.2 shows in summary the approaches and representations used by the two experts, based on model 2, of the developing journey, where the underlying idea is to recreate the movement of the two friends from the beginning until reaching the meeting point (when the distance covered by the two will be 22 km).

Expert A used a tabular representation combined with a schematic representation to illustrate a sequential path that starts at the initial moment and where he signals the elapsed time and the sum of the distances travelled by the two friends over each consecutive hour. In his answer, he explains that he combines a numerical way to get to the answer with a geometrical form of describing the path of the two friends. The expert D only presents a table showing separately the distances travelled by each of the friends and the sum of these distances, over each hour, from the initial moment. He explains that the problem can be dealt with based on the construction of a table. In each of the representations, it can be seen the centrality of the time variation (independent variable) and the co-variation between time and distance.

7.3.2 Definition of Categories

From the analysis of the conceptual models that were present in the five experts' solutions, we concluded that one could basically distinguish two general models on which the solution relies: the completed journey model and the developing journey model. Each of these models could, in turn, reveal different thinking perspectives and result in the use of different mathematical representations.

It would appear from this attempt to categorise the solutions that model 1 is clearly associated with algebraic representations and possibly with a form of textual description, which we may call informal algebra. Although one of the experts have used an $x-y$ graph in one of his solutions (D3), the function of the graph was to illustrate the intersection of two linear functions or, in other words, to illustrate the graphical resolution of a system of two linear equations (space as a function of time).

In the case of model 2, we anticipated the possibility of finding representation modes that are not algebraic nor symbolic, like the use of tables or tabular representations for the sequence of times and distances and the use of diagrams for sketching and expressing abbreviated elements of the problem situation and relations between them. Apparently, model 2 is less likely to be coupled with algebraic and symbolic representations (more precisely, with the formulation and solution of linear equations using the variable time as the unknown).

Thus, we examined the solutions of the students to the given problem, from devising a first set of categories. These preliminary categories were defined as follows (see Table 7.3):

Model 1 (The Completed Journey)

(a) Algebraic/symbolic representation
(b) Textual/descriptive representation

Model 2 (The Developing Journey)

(a) Tables/tabular representation
(b) Diagrammatic/schematic representation

We envisioned that this a priori categorisation could not exhaust all the possibilities of models and forms of representation produced by the students participating in the online competition. At the same time, it was presumed that any of the predefined categories could turn out to be absent in the solutions taken from the competition. Thus, we left the possibility open to include new categories and to eliminate others, as well as the chance to improve and refine the previous categories, based on the actual data.

The encoding of all the participants' solutions was done in three phases. Initially, a researcher carried out a codification of the 220 digital solutions under analysis, noting the distinctive elements of the category assigned to each examined solution: model 1 (a or b) and model 2 (a or b). The solutions that did not fit the predefined categories or raised doubts because they included elements that intersected

Table 7.3 Description of the categories defined in advance to code forms of representation

	Models	
	Model 1	Model 2
Categories defined a priori	*Algebraic/symbolic representations*: use of letters to define variables, use of equations, symbolic expressions, notations and algorithms	*Tables/tabular representations*: tables of numbers, correspondences in tabular format, spreadsheet tables, forms of displaying sequences and counting
	Textual/descriptive representations: use of everyday language, presenting ideas and processes in words, reference to numbers, operations and computations, informal algebra, etc.	*Diagrammatic/schematic representations*: use of diagrams; indicative labels; simplified elements, like segments, points, arrows, letters and brackets; and graphs to represent functions and rules

two categories were coded with the letter X (uncertainty). In a second stage, a second researcher checked the coding already done for the solutions that have not caused uncertainties and validated all encodings that did not differ from the first encoder; the remaining solutions were identified with D (disagreement). In a third phase, the first two researchers and a third researcher examined together, only the solutions designated by X and the solutions designated by D. In the case of the solutions referred to as X, for having raised doubts about the best category to include them, a consensus was reached on the category that could better characterise each solution or, in certain cases, on the need to extend the set with new categories to accommodate those solutions that did not fit any of the previous categories; in the case of the solutions identified with D, the third researcher acted as a decision referee between two competing hypotheses of coding (rate of disagreement of around 1 %).

The process of coding and refinement of the categories proved to be sufficiently flexible and efficient, giving a sign that the initial categories worked reasonably in a large number of solutions. The formulation of new categories and the adjustment of initial encodings were also swift.

7.4 Analysis of the Students' Solutions to the Problem

A global aim of this analysis is to produce a mapping of students' ways of thinking in solving a motion problem in a beyond-school digital environment by identifying the conceptual models that underlie their solutions. Simultaneously, we seek to understand how children express their models with different forms of mathematical representation when using common digital tools and to find out how the use of such digital media is shaping their sense-making of a dynamic situation involving co-variational thinking.

In our data analysis, the unit of analysis was the solving and expressing of a mathematical problem, under the assumption that the digital mathematical discourse (delivered through digital media) is an integral part of the problem-solving activity of the participants in an online competition, in line with the perspective outlined in our theoretical framework (see Chap. 4).

The 220 solutions that were considered correct (which means that the solution was found and the process to achieve it has been submitted) for the motion problem posed on the SUB14 edition of the 2011–2012 competition forms the corpus of data under analysis.

7.4.1 Conceptual Models Involved in the Participants' Problem-Solving and Expressing

The coding of the solutions allowed us to observe that the largely predominant conceptual model in students' approaches was model 2—the developing journey. Thus, there were far fewer solutions supported by model 1—the completed journey.

Only 18 solutions based on model one were found (approximately 8 %). Among them, there are nine solutions that use algebraic representations very close to those expressed by the majority of experts in their more immediate solutions (model 1-a), and the remaining nine solutions matched the type of textual/descriptive representation that arose in a single solution (D4) proposed by one of the experts (model 1-b). Such textual/descriptive representations resemble a quasi-algebraic language, generally comparable to an algebraic approach in establishing relationships between variables, such as equations, but devoid of the usual symbolic formalisation; they rely primarily on verbal language and may include some references to number operations or displaying an informal symbol use. In most of these cases, the participants began by noting that after Alexander had walked 4 km, the two friends would be 18 km away from each other. Then as they approached each other, their relative speed would be 9 km/h (the sum of the two individual speeds); finally, the time elapsed was calculated dividing the distance by the speed. This is similar to the solution D4 that is shown in the summary of the experts' solutions, given in Table 7.1, where no formal use of equations and unknowns is found.

Therefore, the initial category was subsequently refined to better address these representations related to model 1, so as to emphasise the quasi-algebraic language involved. While the textual/descriptive representations submitted by the students suggest nearing an algebraic approach, the solution with similar characteristics by one of the experts was one of several alternatives he found after the most immediate one and therefore seemed to be an attempt to give another "appearance" (perhaps more informal and naïve) to the solution.

Thus, the categories to be considered in model 1 to classify the participants' solutions became the following (see Table 7.4):

Model 1 (The Completed Journey)

(a) Algebraic/symbolic representation (a priori)
(b) Textual/descriptive/quasi-algebraic representation (refined a posteriori)

Regarding model 2, quite a few solutions were obtained that used tabular representations, i.e. model 2-a (51 solutions representing around 23 % of the total), and also a large number of answers that fell into the category of diagrammatic/schematic representations, i.e. model 2-b (44 solutions which correspond to 20 % of the total). However, these two categories previously established for the codification of the solutions based on model 2 were found insufficient to classify all the answers. Consequently, two other categories were created as explained below.

In students' solutions based on model 2, it was observed that the most widely used representation mode was the textual/descriptive representation (87 answers corresponding to approximately 40 % of the total), which had only appeared in the experts' solutions connected to model 1. This suggested the definition of a new category for model 2 characterised by the use of textual/descriptive representations. Clearly, many of the students made their thinking explicit by using everyday language, giving descriptive verbal accounts of the process followed, explaining how they imagined the two friends moving towards each other and recording the distances travelled by them, hour after hour, until the sum of these distances were 22 km. In all cases, the idea is well reflected that the summed distances should make up 22 km and also the fundamental idea that the distance travelled by each of the friends changes with time at constant rates. In many of these solutions, the students chose the variant of separating two stages of the journey: the stage where only Alexander walked and the stage where the two friends walked together in opposite directions. The use of the e-mail window or the text window of the online form available at the competition webpage, without involving any other digital resources, was the most common situation in the solutions submitted by the participants who chose to express their thinking through textual/descriptive representations. The online form offers the facility to perform text formatting, as well as drawings and tables, among other features. Although not comparable to a common text editor, this device proved to be adequate and appropriate for many students to solve and express the problem.

Another set of solutions that emerged from the data with a considerable scale (20 solutions, corresponding to 9 % of the total) contains in a salient form a profusion of pictorial/figurative elements that proved decisive in the expression of the reasoning produced. We thus differentiated two of their essential characteristics for defining a new category in model 2. On the one hand, it is about representations that include a strong visual component reflecting change and coordination as a fundamental part of the expression of the reasoning performed; in other words, the visual elements such as icons, pictures, drawings, colours and other pictorial elements are

Table 7.4 Description of the categories defined subsequently to code forms of representation

	Models	
	Model 1	Model 2
Categories defined a posteriori	*Algebraic/symbolic representations*: use of letters to define variables and use of equations, symbolic expressions, notations and algorithms	*Tables/tabular representations*: tables of numbers, correspondences in tabular format, spreadsheet tables, forms of displaying sequences and counting
	Textual/descriptive/quasi-algebraic representations: use of everyday language, presenting ideas and processes in words, reference to numbers, operations, relations and computations, without symbolic algebra	*Diagrammatic/schematic representations*: use of diagrams; indicative labels; simplified elements like segments, points, arrows, letters and brackets; and graphs to represent functions and rules
		Textual/descriptive representations: use of everyday language, presenting ideas and processes in words, articulating statements in sentences, reference to numbers, operations and computations
		Pictorial/figurative representations: use of drawings, pictures, images and spatial and visual elements like colours, points portraying movement and relative positions over time

7.4 Analysis of the Students' Solutions to the Problem

Table 7.5 Summary of the results of the codification of students' solutions

Model	Category	Solutions Number	(%)	Total
Model 1	Algebraic/symbolic	9	4	18 (8 %)
	Textual/descriptive/quasi-algebraic	9	4	
Model 2	Table/tabular	51	23	202 (92 %)
	Diagrammatic/schematic	44	20	
	Textual/descriptive	87	40	
	Pictorial/figurative	20	9	

not simply an added embellishing element to the solution presented but rather a key aspect to convey the co-variational thinking. On the other hand, these solutions are characterised by visually suggesting a dynamic process (i.e. relative motion) in successively depicting the relative positions of the two friends along the way. To this type of representational character, we attributed the qualification of quasi-dynamical. Therefore, we obtained a new category related to model 2, which is characterised by the use of pictorial/figurative representations.

In brief, two new categories were created to describe the modes of representation used by participants who addressed the problem under model 2, becoming four the number of categories for coding participants' solutions based on this model (see Table 7.4):

Model 2 (The Developing Journey)

(a) Table/tabular representation (a priori)
(b) Diagrammatic/schematic representation (a priori)
(c) Textual/descriptive representation (a posteriori)
(d) Pictorial/figurative representation (a posteriori)

Before we move into a more detailed analysis of the solutions that fall into each category, we present below a table of quantitative results that summarises the data obtained in accordance with the models and categories defined after the second phase of adjustment and category expansion (Table 7.5).

Next, we examine each of the categories through selected prototypical examples in order to better understand the models and associated representations involved in students' approaches to the motion problem with digital tools.

7.4.2 Forms of Representation in Students' Digital Productions

We begin by referring to the slightly refined category used to describe some of the representations used in model 1. This category, related to the textual/descriptive/ quasi-algebraic representations, was visible in some solutions, in which the reasoning

> **Problema nº 1**
> **Resposta**: eles encontraram-se às 11 horas.
> **Resolução:** Eu comecei por pensar que passada uma hora eles estavam a 18 km de distância, porque só o Alexandre tinha andado.
> Depois reparei que passada uma hora com os dois a andar faziam 9 km de distância.
> Dividi os 18 km (distância) por os 9 km que eles andavam numa hora (velocidade) e obtive 2 horas (tempo) que a somar com a que o Alexandre andou sozinho vai dar 3 horas.
> Depois foi só somar às 8 h que foi a hora que o Alexandre abalou e cheguei à conclusão que eles se encontraram às 11 horas.

> **Problem n. 1**
> **Answer**: they met at 11 am.
> **Solution process**: I started by thinking that after one hour they were 18 km away from each other because only Alexander had walked.
> Then I noticed that in each hour, with the two walking, they would make 9 km of distance.
> I divided the 18 km (distance) by the 9 km per hour (speed) and I got 2 hours (time), which when added to the one-hour Alexander had walked alone gives 3 hours.
> Then it was just adding it to the 8 am, which was the time when Alexander left, and I concluded that they met at 11 am.

Fig. 7.2 Example of a solution using a quasi-algebraic type of representation with model 1

resembles a standard algebraic approach, although not containing symbolic algebraic language, i.e. variables and equations, written symbolically. The following are two examples that are included in this category, both making reference to the relevant variables in the problem and explicitly using direct proportionality between the travelled distance and the time.

In the first example (Fig. 7.2), the student realises that after the first hour the distance yet to cover by the two friends is 18 km and grasps the idea that the combined speeds of the two friends mean a relative speed of 9 km/h. Then the ratio between distance and speed is computed to get the elapsed time. There is no indication that the journey undertaken by the two friends was imagined as being in development, which is why it fits a solution based on model 1 using a textual/descriptive/quasi-algebraic type of representation.

The second example shows an analogous reasoning, but this time explicitly referring to the sum of the speeds of the two friends (Fig. 7.3). Like the previous one, it also indicates the division between distance and speed and openly provides other arithmetical operations enabling the finding of the distance travelled by each of the friends, the time each of them has walked and the elapsed time until they met. Here too, there is no use of symbolic algebra or formal equations, but simply natural language and the indication of computations that allow achieving the unknown value.

To complete the model 1-based solutions, we present two examples of answers to the problem that rely on formal algebraic language. In the first example (Fig. 7.4), the participant does not use any type of diagram or scheme and begins by defining variables, assigning them letters. Afterwards, an algebraic equation is formulated and solved to determine the time elapsed until the meeting.

In the second solution presented, which was sent as an attached image file (PNG format) probably created with an image editing software, the participant uses a diagram that represents the route and shows that after 1 h the remaining distance to be covered is 18 km (Fig. 7.5) and then algebraically sets down an equation that allows obtaining the elapsed time (represented by the variable x) and solves it.

7.4 Analysis of the Students' Solutions to the Problem

Resposta:
Os dois amigos encontraram-se às 11 horas. Os dois amigos, em total, tinham de caminhar 22 KM, como o Alexandre sai de casa 1 hora depois do Bernardo, sabemos que só falta 18 KM (22 -4) depois de 1 hora, agora os dois amigos estão a caminhar portanto podemos somar a velocidade dos dois.
4km/h+5km/h=9km/h; 18/9=2; 1+2=3
O Alexandre caminha 12 KM (4x3) em 3 horas, e o Bernardo caminha 10KM (5x2) em duas horas.
10+12=22KM; 8+3=11; 9+2=11

Answer:
The two friends met at 11 am. The two friends, altogether, had to walk 22km; since Alexander leaves home one hour after Bernard, we know that they have only 18km (22 -4) left to walk after the first hour; then the two friends are both walking and therefore we can add their speeds.
4km/h+5km/h=9km/h; 18/9=2; 1+2=3
Alexander walks 12km (4x3) in 3 hours, and Bernard walks 10km (5x2) in two hours.
10+12=22KM; 8+3=11; 9+2=11

Fig. 7.3 Example of another solution using a quasi-algebraic type of representation with model 1

Resposta:
tempo que o Alexandre leva a fazer o percurso= t
tempo que o Bernardo leva a fazer o percurso= t-1
percurso feito pelo Alexandre= 4t
percurso feito pelo Bernardo= 5(t-1)
a soma destes percursos= 22km
4t+5(t-1) =22
4t+5t-5 = 22
9t = 27
t = 3
R: Quando o Alexandre andou 3h e o Bernardo 2h encontraram-se, isto é, às 11h.

Answer:
Time that Alexander takes to make his part of the course= t
Time that Bernard takes to make his part of the course= t-1
Distance covered by Alexander= 4t
Distance covered by Bernard= 5(t-1)
The total of the distances= 22km
4t+5(t-1) = 22
4t+5t-5 = 22
9t = 27
t = 3
Ans.: After Alexander walked 3h and Bernard walked 2h the two met, which means at 11 am.

Fig. 7.4 Example of another solution using an algebraic type of representation with model 1

Turning to the solutions that were coded as belonging to the model 2 approach, our first observations go to the two new categories introduced in the primary list—the textual/descriptive and pictorial/figurative representations—and to their inspection within the empirical data. The textual/descriptive representation is characterised by the use of text, mostly containing verbal and everyday language. It is used by participants as a way to describe their reasoning in words while making the reconstruction of the journey undertaken by the two friends (the moving bodies) over time. Often, this type of solution begins with the presentation of the information given in the problem (different speeds and different times of departure of the two friends). Then it continues with a description of the route undertaken, as time elapses, culminating with the conclusion that the distance of 22 km was covered by the two friends together at exactly 11 a.m.

An illustrative example of a solution in this category is shown in Fig. 7.6.

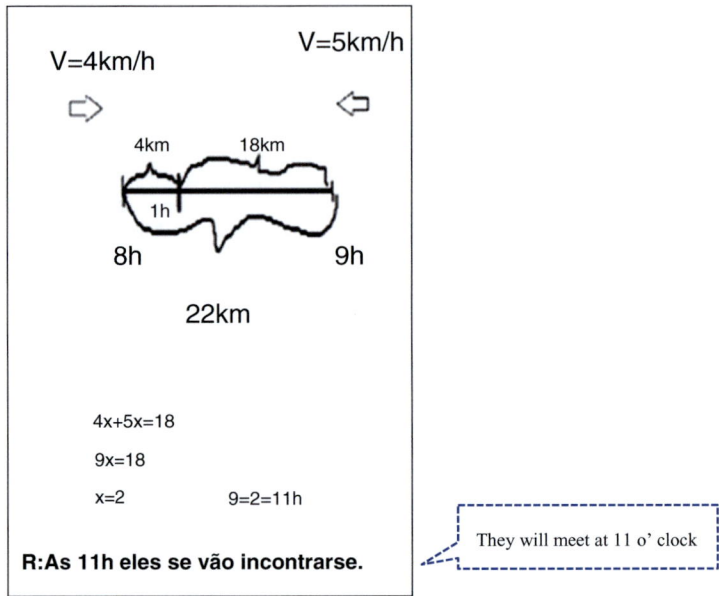

Fig. 7.5 Example of another solution using an algebraic type of representation with model 1 (misspelling in the original answer)

Resposta:	Answer:
Ambos estavam a 22 km separados. O Alexandre estava a uma hora de avanço, logo andou 4km antes de o Bernardo começar, ficando assim a 18 km de distância. Na hora seguinte, o Alexandre andou mais 4 km e o Bernardo partiu e nessa mesma hora andou 5 km, ficando a 9 km de distância. Na hora seguinte, o Alexandre anda mais 4 km e o Bernardo anda mais 5 km, encontran-do-se. Demoraram 3 horas a encontrarem-se, a partir das 8 horas ou 2 horas a partir das 9 horas. Ainda podemos acrescentar que o Alexandre andou 12km e o Bernardo 10km. Encontraram-se às 11 horas.	They were separated by 22km. Alexander was one hour ahead, so he walked 4km before Bernard started to walk, thus becoming 18km apart. In the next hour, Alexander walked over 4km and Bernard started and in that same hour he walked 5km, thus both becoming 9km apart. In the next hour, Alexander walks over another 4km and Bernard walks over 5km more, thus meeting each other. From 8 am it took 3 hours for them to meet, or otherwise from 9 am it took 2 hours. We can also add that Alexander walked 12km and Bernard walked 10km. They met at 11 am.

Fig. 7.6 Example of a solution using a textual/descriptive type of representation with model 2

Another example is also given in Fig. 7.7, where the student actually creates a kind of story to describe the two friends' actions and their developing journey until they met.

Another solution sent as an attachment in JPEG format, which contains an extensive and detailed description of the reasoning produced is given below (Fig. 7.8). Although the picture sent does not contain any kind of diagram or picture, it attempts

7.4 Analysis of the Students' Solutions to the Problem

Resposta:	Answer:
Resposta ao problema 1	Solution to problem 1
O Alexandre sai de casa às 8 horas da manhã e vai andando a 4 km por hora. Chegam as 9 horas da manhã e o Alexandre já percorreu 4 km e o Bernardo acaba de sair de casa e vai andando a 5 km por hora. Chegam as 10 horas da manhã e o Alexandre já percorreu 8 km e o Bernardo 5 km e 8 + 5 = 13 e como a distância entre as casas deles é de 22 km ainda não se encontraram. Chegam as 11 horas da manhã e o Alexandre já percorreu 12 km e o Bernardo 10 km e 12 + 10 = 22 e como a distância entre as casas deles é de 22 km já se conseguiram encontrar e isto aconteceu às 11 horas da manhã.	Alexander leaves home at 8 o'clock in the morning and goes walking at 4 km per hour. When it's 9 o'clock in the morning and Alexander has already travelled 4 km Bernard just leaves home and goes walking at 5 km per hour. At 10 o'clock in the morning Alexander has already travelled 8 km and Bernardo 5 km, and knowing that 8 + 5 = 13 and as the distance between their homes is 22 km, they still haven't met. At 11 o'clock in the morning Alexander has already travelled 12 km and Bernardo 10 km, and knowing that 12 + 10 = 22 and as the distance between their homes is 22 km they finally managed to meet and that happened at 11 o'clock.
R: Eram 11 horas.	Ans.: It was at 11 o'clock.

Fig. 7.7 Example of another solution using a textual/descriptive type of representation with model 2

to make an initial organisation of the information into two zones, one on the path of Alexander (left side) and another on the path of Bernard (right side). There follows an analysis of the relative positions of the two friends and, finally, the answer.

A fourth solution belonging to this category shows that the textual/descriptive representations include, in many cases, more abbreviated, schematic and synthetic forms of expressing the reasoning developed. In the following figure, it may be observed that there is an abbreviated indication of the sum of the distances travelled by the two friends, over each passing hour (Fig. 7.9).

With regard to the other category that has emerged from the data, which we have named pictorial/figurative, one of its most obvious features is the ability to transmit a dynamic situation through a static figure, either through the use of colours, icons, drawings or images. It clearly projects the visual nature of the representations used to deal with the motion problem.

A good example of a pictorial/figurative solution is shown in Fig. 7.10. The solution presented by this participant was sent as a PowerPoint file attachment. The straight path between the two friends' homes is represented by a line segment divided into intervals of 2 km, and the small circles of two different colours show their positions, relative to one another and to each of their respective homes, every half hour. The variation of time is recorded on the left margin, showing that each of the positions of the two circles is corresponding to each of the successive instants. The two overlapping circles show the meeting point, and the final answer is given in a short phrase: "They will meet at 11 a.m.".

Another solution that has the same type of characteristics, namely, the quasi-dynamic aspect, is shown in Fig. 7.11. This solution was also sent as a file attachment, this time in a Word document that includes several objects constructed with

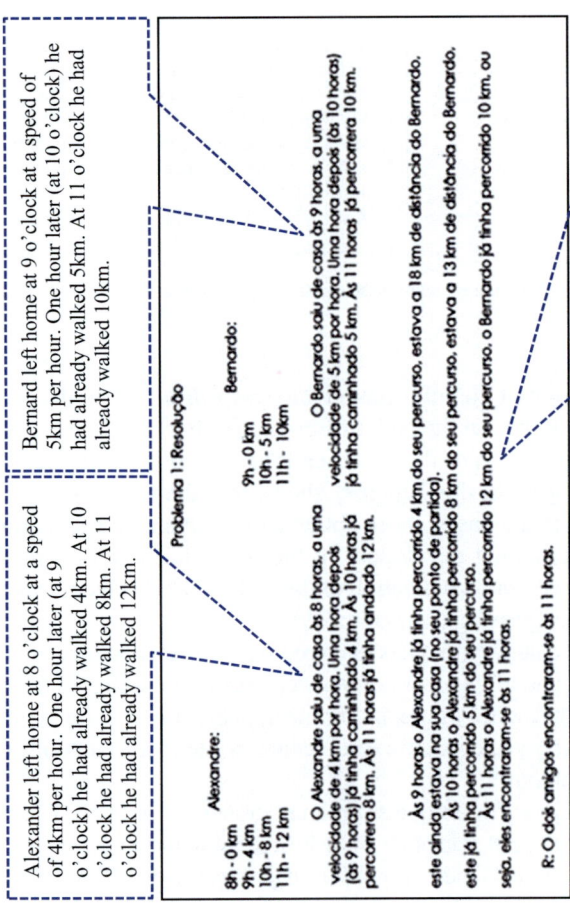

Fig. 7.8 Example of another solution using a textual/descriptive type of representation with model 2

7.4 Analysis of the Students' Solutions to the Problem

Resposta:

Dados:
O Bernardo e o Alexandre moram a 22km um do outro
Alexandre: Abalou às 8h de casa e anda 4km/h
Bernardo: Abalou às 9h de casa e anda 5km/h

Resolução:
Às 9h o Alexandre tinha andado 4km e o Bernardo 0km -> 4km+0km=4km
Às 10h o Alexandre tinha andado 8km e o Bernardo 5km -> 8km+5km=13km
Às 11h o Alexandre tinha andado 12km e o Bernardo 10km -> 12km+10km=22km
Resposta: Claro, às 11h o Alexandre e o Bernardo vão-se encontrar, porque percorreram os 22km que havia entre as suas casas.

Answer:

Givens:
Bernard and Alexander live 22km away from each other
Alexander: Left at 8h from home and walks at 4km/h
Bernard: Left at 9h from home and walks at 5km/h

Solution process:
At 9h Alexander had walked 4km and Bernard 0km -> 4km+0km=4km
At 10h Alexander had walked 8km and Bernardo 5km -> 8km+5km=13km
At 11h Alexander had walked 12km and Bernard 10km -> 12km+10km=22km
Answer: Of course, at 11h Alexander and Bernard will meet because they have undertaken the 22km between their homes.

Fig. 7.9 Example of another solution using a textual/descriptive type of representation with model 2

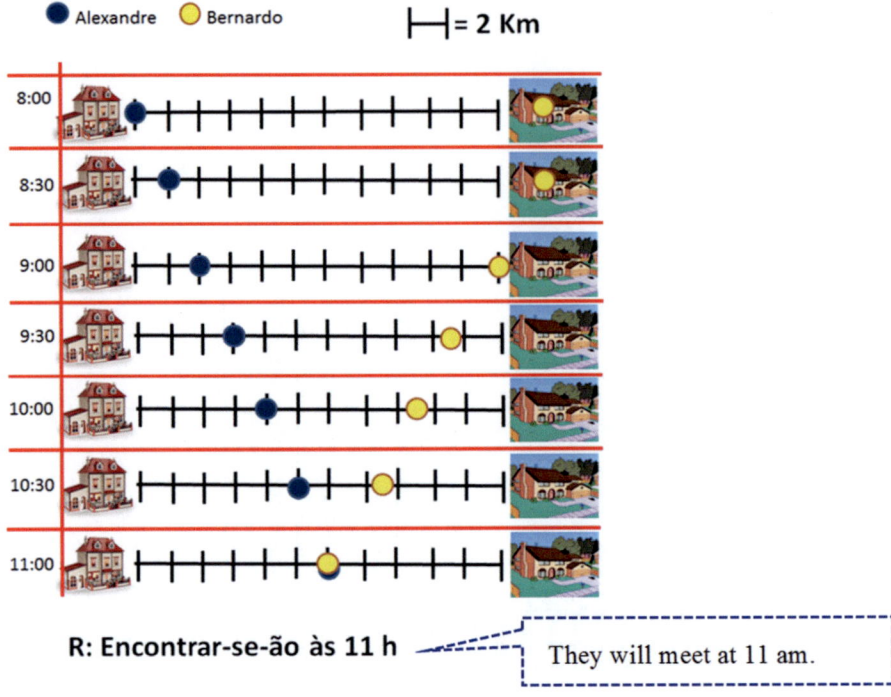

Fig. 7.10 Example of a solution using a pictorial type of representation with model 2

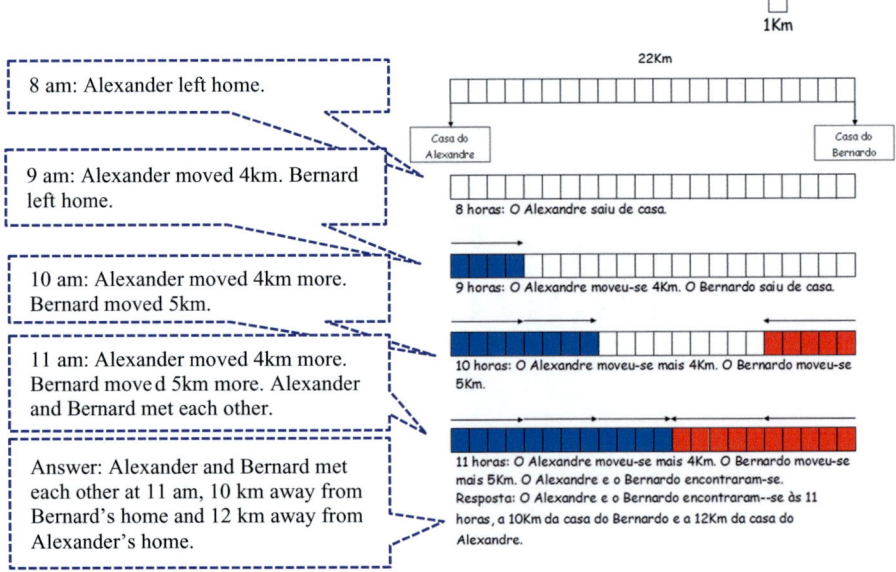

Fig. 7.11 Example of another solution using a pictorial type of representation with model 2

7.4 Analysis of the Students' Solutions to the Problem

the tools that let you insert, set and format shapes. The small squared cells are obtained by using a table (with a single row and 22 columns) that works as a corridor of squares, which are successively appearing shaded through the formatting of the background colour of the cells. In this case, the use of colour is important, because filling the cells with distinct colours on each side of the path both shows the relative position of the two friends over time, the distance that they each have already walked and also the distance that still separates them. This representation uses units of 1 km to show the progression along the route and uses the time change in intervals of 1 h.

Some of the facets that are quite prominent in this category of solutions are the use of colours and of small textual captions and notes, as well as a clear choice of use of pictures that enable the creation of an enlightening picture of the situation to be modelled. As with the solution presented above in Fig. 7.10 (which uses the circles of two colours), the following solution was also produced in a PowerPoint file and also introduces images to recreate the homes of the two friends and to spot the positions where each friend starts walking (Fig. 7.12). In this third solution, there is again the presence of pictures that illustrate the two friends. Moreover, different lengths of arrows are used (for the progress of Alexander and the progress of Bernard), indicating the different distances travelled by each one, hour after hour, combined with labelling in text captions. The pairs of arrows suggesting the two friends simultaneously walking are also signalled, showing that each pair of arrows is relative to a time period of 1 h.

In any of these solutions, that visually reflect model 2, it is quite clear how the digital representations are effective means of realising the idea of a developing journey from the initial moment until the moment of the meeting, i.e. giving a picture of a conceptual model of co-variation between the distance travelled and time and also between the displacement of each friend and time.

Next, some examples of tables/tabular representations will be shown, which constituted the second largest category of representations used in the solutions that were supported on the model 2. We considered the common layout of tables (rows and columns) but also other forms of representation showing a tabular organisation though not clearly displaying the separation of the lines and columns.

In several cases, as in the solution pictured in Fig. 7.13, the tables are accompanied by explanatory text on how they were constructed. The solution from this participant was sent as an image (PNG format).

The solution in Fig. 7.14 was submitted using the online form available on the competition website that provides a table editor. In this solution, the table contains a column for the cumulative distance covered by the two friends together, under the heading of "total of kilometres covered".

Figures 7.15 and 7.16 illustrate two other representations of a tabular form: the first is obtained by simple tabulations of the text made in the online form window (Fig. 7.15); the second shows the construction of two lists of pairs of values (time-displacement covariance), both working in conjunction to verify if the total distance is covered at a given moment, as it is highlighted in bold (Fig. 7.16). While the first list of pairs shows the progress of Alexander's journey relative to his initial position,

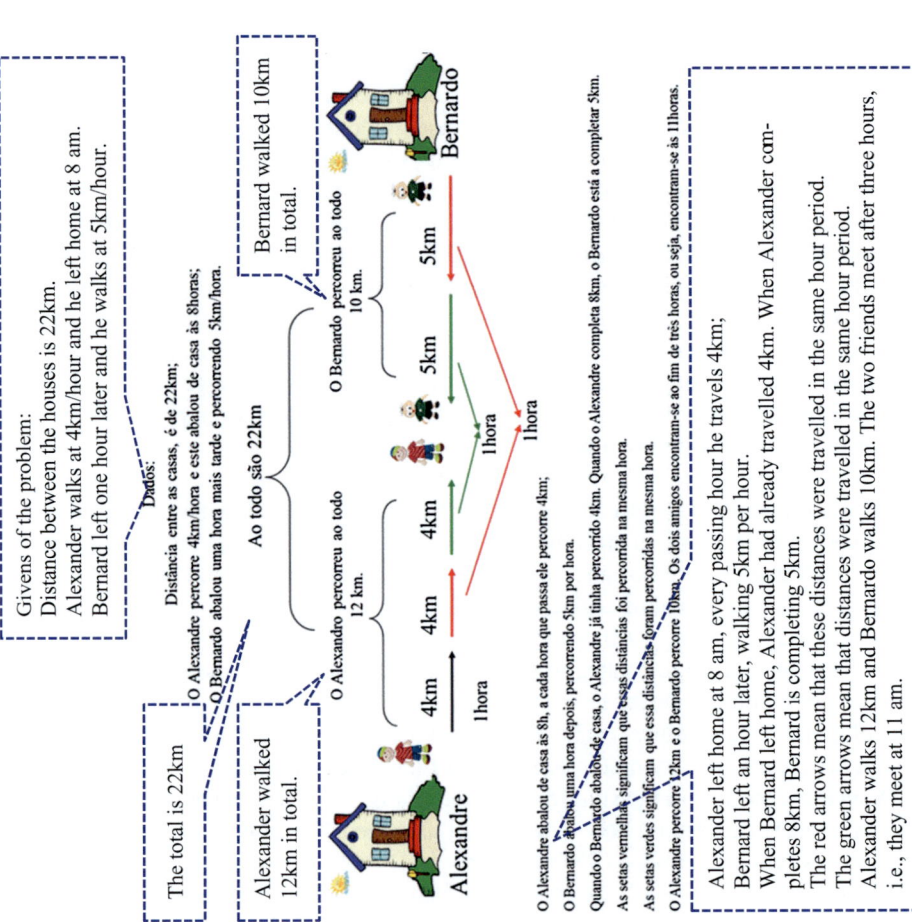

Fig. 7.12 Another example of pictorial/figurative representation with model 2 created in PowerPoint

7.4 Analysis of the Students' Solutions to the Problem

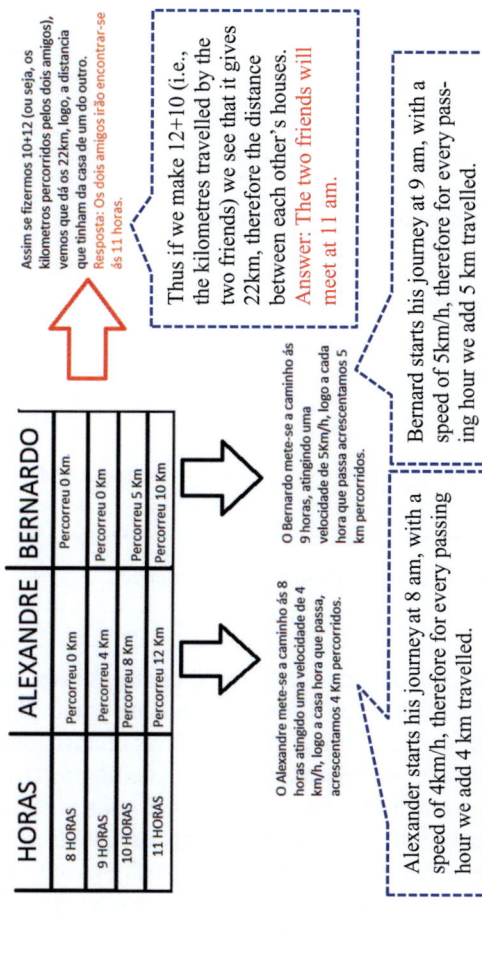

Fig. 7.13 An example of tabular representation with model 2

Casa Alexandre		22km	Casa Bernardo
08:00h			09:00h
4Km/h			5Km/h

Horas	Alexandre - km percorridos	Bernardo - km percorridos	Total de km percorridos
08:00	0	0	0
09:00	4	0	4
10:00	8	5	13
11:00	12	10	22

Resposta: Eles encontram-se ao 11:00h, porque a esta hora o total de km percorridos foi de 22km.

Fig. 7.14 An example of tabular representation with model 2

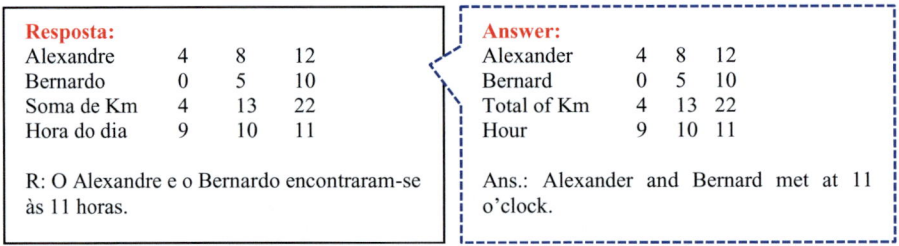

Resposta:
Alexandre   4   8   12
Bernardo   0   5   10
Soma de Km   4   13   22
Hora do dia 9   10   11

R: O Alexandre e o Bernardo encontraram-se às 11 horas.

Answer:
Alexander   4   8   12
Bernard   0   5   10
Total of Km 4   13   22
Hour   9   10   11

Ans.: Alexander and Bernard met at 11 o'clock.

Fig. 7.15 Another example of tabular representation with model 2

the second list indicates the progress of Bernard relative to his initial position and additionally his position relative to Alexander's starting point. This example points out that both friends are walking in opposite directions relative to a reference frame: Bernard's travelled distance is translated into his position relative to Alexander's point of departure, which entails that the distance to that point is decreasing (Fig. 7.16).

In the case of the solutions that presented diagrammatic/schematic representations, it is clear that how the diagrams produced disclosed the underlying conceptual model. In the figures that follow, participants' productions are similar to the kind of diagrammatic representation proposed in the solution A3 of expert A. Typical of this kind of representation are elements like lines, points, dashes, indicative labels, use of simple shapes, arrows, curly brackets, letters and numerical information (Figs. 7.17, 7.18 and 7.19). It should be recalled that this was the third most abundant category in the solutions that were based on model 2, immediately after the category denoting the table/tabular representation. This is, obviously, a strong contrast with the rare presence of such diagrams in the experts' solutions. Students' co-variational reasoning is apparently not only linked to a conceptual model that

7.4 Analysis of the Students' Solutions to the Problem

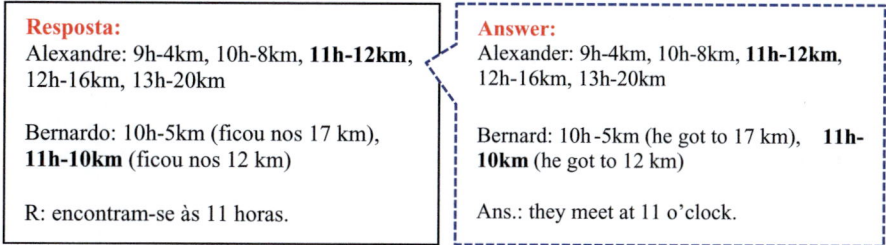

Fig. 7.16 Another example of tabular representation with model 2

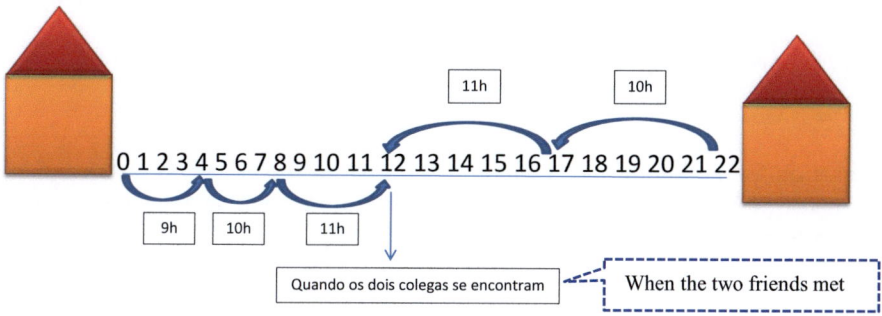

Fig. 7.17 An example of diagrammatic representation with model 2

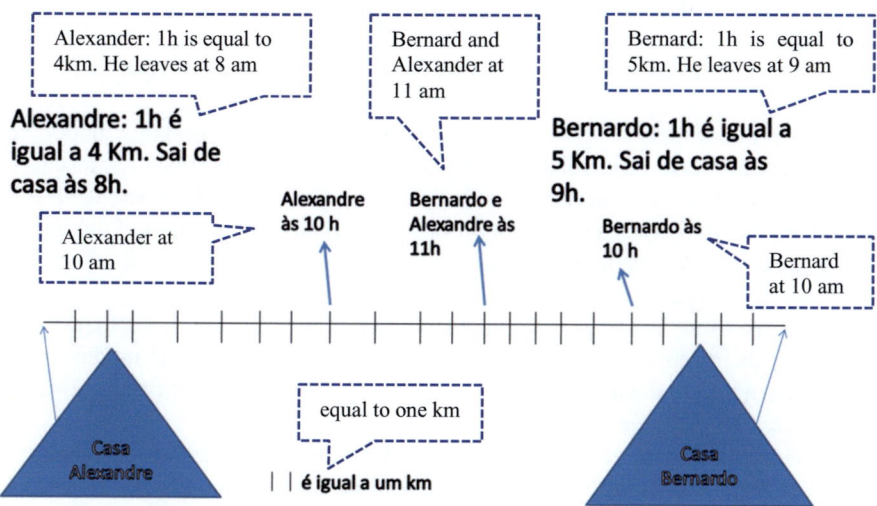

Fig. 7.18 Another example of diagrammatic representation with model 2

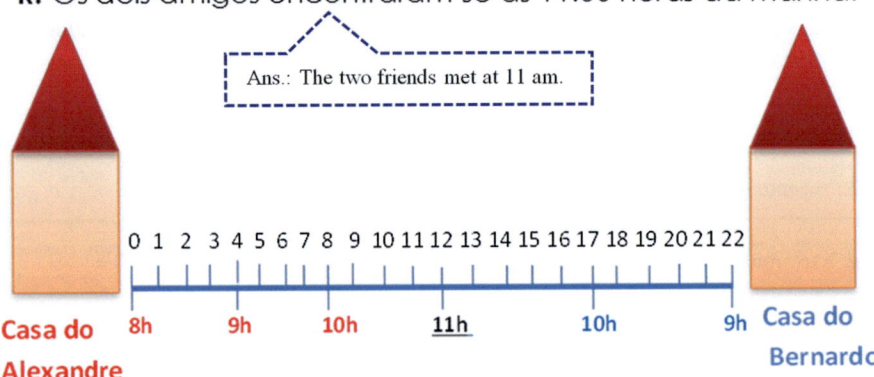

Fig. 7.19 Another example of diagrammatic representation with model 2

stands on the notion of a developing motion but also it is largely expressed by means of tables (sequentially ordered numerical information) and diagrams revealing sequentially organised visual information.

7.5 Discussion and Conclusion

We set ourselves the task of scrutinising the approaches used by a high number of participants in SUB14 to a motion problem (all solutions submitted by 8th graders). Our effort concerned not only understanding their conceptual models but also looking at the ways in which their uses of digital technologies were associated with such models.

Our results strongly support the view that problem-driven conceptual models are powerful windows to make sense of how young people put into action their resources and their abilities to reason mathematically in effective ways. Moreover, the postulated conjunction between solving a problem and expressing the underlying model of the problem situation appears evident from the results of the analysis of the 220 solutions considered.

As shown from the categories defined to describe the different models and forms of representation that were associated to them, the same conceptual model can be expressed in different ways and can appeal to quite distinct types of inscriptions and communication vehicles.

At the same time, it was possible to see how young people can take advantage of common usage digital tools, with which they seem comfortable and accustomed to deal, for serving their attempts of getting a convenient way to express their mathematical models.

7.5 Discussion and Conclusion

The proposed problem is an opportunity to understand the participants' ability (12–14-year-olds) to engage in co-variational reasoning in a situation of relative motion. Although these students had previous school learning in algebra and algebraic procedures (such as solving linear equations), the data revealed that their use of formal algebra was very sparse. The interesting thing is that this appears to be closely associated with the fact that they have largely developed a model that, unlike the more common model in experts' solutions, is not very compatible with the algebraic symbolic methods to describe the situation. This is not completely unknown to research. In fact, Greeno and Hall (1997) observed that more than half of the subjects working on a motion problem (students, teachers and engineering undergraduates) tried representation systems other than algebraic expressions.

> The representational work that people do often uses nonstandard forms, which are constructed for the immediate purpose of developing their understanding. In most practices, people generate representational forms in ways that serve immediate local purposes. In addition to being representations *of* something, they are *for* something. (Greeno & Hall, 1997, p. 365)

In our participants' approaches, we could distinguish two main conceptualisations: one is thinking of the journey undertaken by the two friends as being completed and the meeting having been consummated (model 1) and the other is imagining the journey as being in progress with the two friends changing their relative positions over time until they will eventually meet (model 2). We have proposed two opposite ideas to characterise the two models, being the first one described as looking at the past (one needs to find out the elapsed time) and the second one as looking to the future (one needs to reconstruct sequentially the relative motion). The coding of the solutions collected showed a strong prevalence of the model 2, amounting to 92 % of the answers pertaining to this model category.

This overall result provides a clear indication of how the participants' co-variational reasoning was generally produced and expressed. The simultaneous change of displacement and time was strongly associated with the idea of continuous evolving change. Model 2 reflects such continuous change in an unequivocal way and shows in a number of different ways how the co-varying variables persisted in students' images of the situation. As Carlson et al. (2002) have argued, this phenomenon is at the core of co-variational reasoning, and it comes down to the fundamental issue of coordination.

We may thus conclude that the participants' models were essentially focused on coordinating and maintaining control over the simultaneous variation of time and space: either through verbal narratives in reconstructing the journey, through tables, through diagrams or via figurative visual registers, many of them displaying a pictorial sequence of frames.

In a way, this result suggests that students' non-formal and non-algorithmic mathematisation of the relative motion is grounded in a seeming perception of the important role played by coordination and this has an effect on the kind of adaptive interpretations they have developed, in line with the concept referred to by Izsák and Findell (2005). Evidence suggests that these youngsters engaged in various adaptive interpretations that drove them away from the classical algorithmic perspective of focusing on a specific value of a variable, that is to say, finding the unknown (the algebraic modelling of the dynamic functional event), to using other representational forms, which supported their understanding and visualisation of change over time.

Furthermore, the data also indicated how the distribution of different forms of representation was associated with participants' prevalent model. Four subcategories were obtained in connection to model 2, giving strong support to the idea that the textual/descriptive representations were important tools for a large number of solvers (Passaro, 2009). The diversity shown in youngsters' creative representations comes as a huge contrast to the relatively few forms of representation suggested by the five experts who collaborated in the study.

When we consider the fact that for many participants the narrative, the language of everyday life and the explanation based on words and number operations represented an appropriate way to express their conceptual model, then we have reason to believe that the most common digital tools available on a computer work well to solve and express the given problem. Thus, just resorting to a text editor or to a text online window becomes a good choice for many. The same is true if we consider the use of tables and tabular representations. It is apparent in our data that a large number of participants are competent in using, creating, formatting and drawing tables or just arranging information in a tabular disposition. And this was manifestly another important tool to express and solve the problem of the two friends walking towards each other.

Finally, the graphical character of a high percentage of answers to the problem was very relevant. In addition to the representations of a more schematic or diagrammatic nature, we must emphasise the figurative/pictorial representations that are indicative of a digital enactment of the problem. Drawing on Sherin's (2000) view of the value of students' invented ways of representing motion and of Monaghan and Clement's (2000) considerations about enactment and self-projecting in conceptualising relative motion, we may be able to suggest that such particular solutions represent a digitally expository discourse that not only included enactment and a sense of physical movement but also a convincing image of co-variation in a dynamic situation. The forms of representation involved are consistent with other studies that also detected the drawing of successive states in motion, the use of a table that organises intermediate quantities during those states or oral narratives that coordinated time and distance across the motion trajectory (Greeno & Hall, 1997).

We found out that various digital tools supported this kind of highly visual expression of relative motion, as in the case of text editor, the presentation editor, the drawing software and in general all means of creating multimedia documents. It is has to do with using pictures and other digital objects, many of them displaying a pictorial sequence of frames that present themselves full of dynamism, colour, detail, realism and offer immediate insight and reading. We claim that this particular solving and expressing of a motion problem is in line with the concept of visuality developed by Presmeg (2006) and is consistent with some of the insightful "ways in which drawings are adapted to deal with the requirements of representing motion"

(Sherin, 2000, p. 416). It is, in this case, a distinctive visuality, consistent with the multimedia language that digital technologies infuse in our world and undoubtedly in young people's world.

This digital visuality is a significant trait of the pictorial/figurative representations observed in a particular subcategory of the solutions based on model 2. Those represent prototypical instances to envisage how, in the case of solving a motion problem, young participants in the competition are performing as humans-with-media (Borba & Villarreal, 2005) in solving and expressing their mathematical reasoning and in developing concepts which are rooted in their common sense reasoning (Doorman & Gravemeijer, 2008). In any case, even the non-figurative, the textual or the tabular representation forms also appear interconnected with a variety of digital home technologies and give a coherent view of how young children make use of technology in dealing with moderate challenging motion problems.

Finally, in future studies, it will be possible to extend the analysis to other motion problems which have been proposed over several editions of the competitions. It must be pointed out that the problem discussed here may be seen as limited, especially due to the fact that it only requires working with whole numbers. Therefore, it would be reasonable to hypothesise that the model of the developing journey would not be as effective if the solution were a decimal number or a fraction.

In addition to consider other versions of motion problems, it will be interesting to extend the analysis to the solutions that were produced by the young competitors in the Final, where they only used paper and pencil. In terms of research development, a step forward might be to compare the conceptual models involved in solving a motion problem with and without digital tools.

References

Borba, M. C., & Villarreal, M. E. (2005). *Humans-with-media and the reorganization of mathematical thinking: Information and communication technologies, modeling, visualization and experimentation*. New York, NY: Springer.

Carlson, M. (2002). Physical enactment: A powerful representational tool for understanding the nature of covarying relationships? In F. Hitt (Ed.), *Representations and mathematics visualization* (pp. 63–77). Special Issue of PME-NA & Cinvestav-IPN.

Carlson, M., Jacobs, S., Coe, E., Larsen, S., & Hsu, E. (2002). Applying covariational reasoning while modeling dynamic events. *Journal for Research in Mathematics Education, 33*(5), 352–378.

Carlson, M., Larsen, S., & Lesh, R. (2003). Integrating a models and modeling perspective with existing research and practice. In R. Lesh & H. Doerr (Eds.), *Beyond constructivism – models and modeling perspectives on mathematics problem solving, learning, and teaching* (pp. 465–478). Mahwah, NJ: Erlbaum Associates.

Carraher, D. W., Martinez, M. V., & Schliemann, A. D. (2007). Early algebra and mathematical generalization. *ZDM: The International Journal on Mathematics Education, 40*(1), 3–22. doi:10.1007/s11858-007-0067-7.

Doorman, L. M., & Gravemeijer, K. P. E. (2008). Emergent modeling: Discrete graphs to support the understanding of change and velocity. *ZDM, 41*(1–2), 199–211. doi:10.1007/s11858-008-0130-z.

Greeno, J. G., & Hall, R. P. (1997). Practicing representation: Learning with and about representational forms. *Phi Delta Kappan, 78*(5), 361–367.

Izsák, A., & Findell, B. R. (2005). Adaptive interpretation: Building continuity between students' experiences solving problems in arithmetic and in algebra. *ZDM: The International Journal on Mathematics Education, 37*(1), 60–67. doi:10.1007/BF02655898.

Matos, J., & Carreira, S. (1997). The quest for meaning in students' mathematical modeling. In K. Houston, W. Blum, I. Huntley, & N. Neill (Eds.), *Teaching and learning mathematical modeling – ICTMA 7* (pp. 63–75). Chichester, UK: Horwood Publishing.

Monaghan, J. M., & Clement, J. (2000). Algorithms, visualization, and mental models: High school students' interactions with a relative motion simulation. *Journal of Science Education and Technology, 9*(4), 311–325.

Passaro, V. (2009). Obstacles à l'acquisition du concept de covariation et l'introduction de la représentation graphique en deuxième secondaire. *Annales de Didactique et de Sciences Cognitives, 14*, 61–77.

Presmeg, N. C. (2006). Research on visualization in learning and teaching mathematics. In A. Gutiérrez & P. Boero (Eds.), *Handbook of research on the psychology of mathematics education: Past, present and future* (pp. 205–235). Rotterdam, The Netherlands: Sense Publishers.

Santos-Trigo, M. (1996). An exploration of strategies used by students to solve problems with multiple ways of solution. *Journal of Mathematical Behavior, 15*(3), 263–284. doi:10.1016/S0732-3123(96)90006-1.

Santos-Trigo, M. (2004). The role of technology in students' conceptual constructions in a sample case of problem solving. *Focus on Learning Problems in Mathematics, 26*(2), 1–17.

Sherin, B. L. (2000). How students invent representations of motion: A genetic account. *Journal of Mathematical Behavior, 19*, 399–441.

Zeytun, A., Çetinkaya, B., & Erbaş, A. (2010). Mathematics teachers' covariational reasoning levels and predictions about students' covariational reasoning. *Educational Sciences: Theory and Practice, 10*(3), 1601–1612.

Chapter 8
Youngsters Solving Mathematical Problems with Technology: Summary and Implications

Abstract The final chapter summarises the overall findings of the Problem@Web project and considers the implications of the findings in terms of how the youngsters of today tackle mathematical problems and communicate their mathematical problem-solving. With data from the youngsters' participation in two online mathematical problem-solving competitions that were characterised by moderately challenging problems, we found that the youngsters we studied had domain over a set of general-use digital tools and while they were less aware of digital resources with a stronger association with mathematics they were able to gain many capabilities by tackling the mathematical problems and seeking expeditious, appropriate and productive ways of expressing their mathematical thinking. In this respect, they were able to harness their technological skills while simultaneously developing and improving their capacity to create and use a range of mathematical representations. We explain this as co-action between the tool and the solver, with this interconnectedness leading to jointly developed technological skills and mathematical skills that result in the capacity of mathematical problem-solving with technology. Given the possibility of youngsters developing this capacity, a key issue is how this can be harnessed to promote the success of youngsters in mathematics in our digital era.

Keywords Co-action • Digital-mathematical discourse • Digital tools • Humans-with-media • Mathematical problem-solving and expressing • Mathematics teachers' involvement • Moderate-challenging problems • Problem@Web project • Students' technological competencies • Web-based competitions

8.1 Introduction

There continues to be much discussion about the uses made of technologies by youngsters born into the current digital era. Some express worries about the amount of time today's youngsters might be spending using digital technologies; others point to young people learning much through their everyday use of digital technologies. School systems across the globe have, over recent years, sought to address the challenges raised by the increasing availability of various digital technologies. As a result, many countries have been making efforts to equip schools with ever-better technologies. At the same time, the availability of digital technologies at home has been growing to such an extent that youngsters' home access to new projects and initiatives taking place through the Internet is superior to their school access.

This growing dissemination of digital technologies accentuates the necessity of investigating how today's youngsters find effective and productive ways of thinking about mathematical problems and how they might achieve a solution and communicate it using any digital resources that they have at hand. While there has been something of a decline in more traditionally orientated research on students' mathematical problem-solving (English, 2008; English & Sriraman, 2010; Lesh & Zawojewski, 2007), several international trends, many connected with the OECD's Programme for International Student Assessment (PISA), are repositioning problem-solving in a privileged position in mathematics learning (Anderson, Chiu & Yore, 2010; Törner, Schoenfeld & Reiss, 2007).

All this means that there is still much to learn about the connection between the use of technologies and the process of solving mathematical problems. That is what the Problem@Web project sought to investigate. In this project, our main research topics were the youngsters' types of problem-solving approaches (such as visualisation, experimentation and simulation) through the use of digital technologies and the different types of mathematical representations available and afforded by technologies that youngsters are creating to express and carry out mathematical thinking.

As we argued in Chap. 1 of this book, a major challenge remains to reveal the impact of the widespread accessibility of digital technologies on young people's capability to use them for learning and for problem-solving. In this chapter, we summarise the contribution to understanding youngsters solving mathematical problems with technology that we have made through the Problem@Web project. In this project, we analysed the way in which today's young people tackled mathematical problems with the technologies of their choice when having an extended time for finding and communicating their solutions.

In what follows, we provide a brief outline of the Problem@Web project and the methodology we used for the project. We then examine the youngsters who participated and the views of their teachers. We then outline the theoretical stance that we developed for the project and summarise the findings in terms of the major mathematical concepts of invariance, quantity variation and co-variation. Following a discussion of the findings, we conclude with the implications of our project and suggestions for further research.

8.2 The Problem@Web Project

Our research project, *Mathematical Problem Solving: Perspectives on an interactive web-based competition*, which we call Problem@Web, grew out of our interest in understanding how the youngsters of today tackle and solve moderately challenging mathematical problems using the digital tools of their choice. This interest was fuelled by our experience of two online mathematical problem-solving

SUB12-Problema 10
A jogar ao berlinde

O Pedro, o David e a Joana estão a jogar ao berlinde. Ao todo, os três amigos têm 198 berlindes. O Pedro tem 3 vezes mais berlindes do que a Joana e o David tem 2 vezes menos berlindes do que a Joana.
Quantos berlindes tem cada um deles?

Não te esqueças de explicar o teu processo de resolução.

Data limite de envio de respostas ao PROBLEMA 10:
Domingo, 31 de Maio.

Pedro, David and Joana are playing with marbles. Altogether, the three friends have 198 marbles. Pedro has 3 times more marbles than Joana and David has 2 times less marbles than Joana.
How many marbles has each of the friends?

Do not forget to explain your problem solving process.

Fig. 8.1 Image of Problem #10 of SUB12 (edition 2010–2011) as it was posted on the webpage and of the typical buttons for opening the digital form and for downloading the problem

competitions, called SUB12 (aimed at 10–12-year-olds) and SUB14 (aimed at 13–14-year-olds), and the youngsters who chose to participate in them (Fig. 8.1).

From the very beginnings of the competitions in 2005, the solutions submitted by the young participants showed a wealth of ways of representing and expressing mathematical reasoning. We were intrigued by the evidence of innovative ways of mathematical thinking and problem-solving by youngsters who engaged in mathematical thinking via the computer in order to participate in the online competitions. There were digital images, textual written compositions, diagrams, tables and graphs using various file formats including Word, Excel, GSP and GeoGebra (Fig. 8.2). This inspired us to try to understand their approaches to mathematical problem-solving.

♦ Pedro = 3 Joanas
♦ Se o David tem metade dos berlindes da Joana, então a Joana tem o dobro dos berlindes do David, então:
JOANA = 2 DAVID
♦ Se o Pedro tem o triplo dos berlindes da Joana, então tem o sêxtuplo dos berlindes do David, então:
PEDRO = 6 DAVID

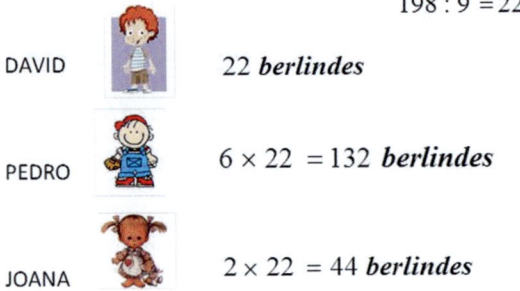

$$198 : 9 = 22$$

DAVID   22 *berlindes*

PEDRO $6 \times 22 = 132$ *berlindes*

JOANA $2 \times 22 = 44$ *berlindes*

Fig. 8.2 Resolution of Problem #10 of SUB12 (edition 2010–2011) sent in a Word file showing a written composition combined with pictures and a small table exhibiting the ratios between the three children and the calculations done to find the three quantities of marbles

Our Problem@Web project had three main research foci:

(a) The youngsters' strategies for solving mathematical problems, modes of representation and expression of mathematical thinking and the use of digital technologies in problem-solving
(b) The youngsters' attitudes and emotions related to mathematics and mathematical problem-solving, both in school and beyond-school activities, considering students, parents and teachers
(c) The youngsters' creativity expressed in their mathematical problem-solving and its relation to the use of digital technologies

In this book, we focus on the first of these foci. In this, a main sources of data for us were the solutions to the problems that we received over the years. This resulted in a huge and varied database. We also collected all the emails and answers submitted throughout three editions of the competitions, together with other informative data. Furthermore, we interviewed many young participants, as well as some former participants, several mathematics teachers who followed the competitions and also youngsters' parents and relatives.

In addition to this qualitative methodology, we conducted an online survey of participating youngsters in the 5th to 8th grades from the Algarve region in southern Portugal. The response rate was close to 20 %, having obtained 350 answers. The questionnaire focused on the youngsters' relationship with technologies, their relationship with mathematics and problem-solving and their views on participating in the SUB12 or SUB14 online mathematics competitions.

As we were aware that the mathematical problems published on the competitions' website were widely used in schools, both in mathematics classes and in other school contexts such as libraries, mathematics after-school clubs or tutorial classes (also known as supervised study classes), this provided the opportunity to extend our research to the school context. As a result, we complemented the information gathered in the online and beyond-school competitions with information from classrooms, thereby allowing us to develop a more rounded understanding of the phenomenon of youngsters solving non-routine mathematical problems with technology in the school context. This meant, for example, that we could obtain data from an 8th grade class (of 13–14-year-olds) where students spent a number of lessons tackling various problems given in the SUB14 competition.

Given this very large corpus of data, it was necessary to establish some criteria to select the data to be analysed. One of these criteria was the possibility of comparing students' productions in the competition with their efforts in the school context. To develop our data analysis, we chose problems from different mathematical topics to, in a sense, be considered as a trilogy as they address the major concepts of invariance, quantity variation and co-variation. This choice allowed us to give a very broad idea of the wealth of possibilities that technology use can provide in problem-solving and in problem-driven conceptual development.

The analysis of the selected elements of data utilised both data-driven encoding (a form of pattern recognition within the data where emerging themes become the categories for analysis) and theory-driven coding (where a preliminary set of categories are developed a priori based on the research question and the theoretical framework). For example, looking for patterns and themes from a set of solutions to a problem involving quantity variation led to ways of ranking different spreadsheet-based solutions revealed in those solutions; in addition, we used theoretical concepts, like the idea of *co-action* between the solver and the tool to complement the first encoding and generate evidence of the relationship between the resolution of the problem and the use of the spreadsheet, thus reaching a second level of interpretive understanding (Fereday & Muir-Cochrane, 2006). This is how, by looking at our data from new and different angles, we discovered new aspects and new ways to question and discuss the data. It is through this process that we have sought to generate analytical theory and extend pre-existing theory from our data.

In the next section, we provide a characterisation of the youngsters who participated in the SUB12 or SUB14 online mathematics competitions and whose submitted solutions are the basis of the analysis presented in subsequent sections.

8.3 The Youngsters Solving Mathematical Problems with Technology

Throughout the various editions of SUB12 and SUB14 competitions, we have realised that the participants resorted to various technological tools of daily use, commonly available on personal computers, such as text editors or image editors and spreadsheets, among others. In any case, over the years, there were always submissions that the participants had done with paper and pencil and then scanned or photographed and the files sent as attachments. Likewise, not all young people sent files as attachments as some preferred to use the text editing tool embedded in the provided online form to present their solution and to describe the process of solving the problem.

Such plurality, together with the fact that the competitions take place through the Internet and require a digital and asynchronous communication by e-mail or through the website, motivated a strong interest in knowing in more detail who are the youngsters that engage in these competitions and what kind of relationship they have with digital technologies, especially in terms of what they are capable of doing with them when dealing with mathematical problems and communicating mathematically.

With this aim, our option for gathering data was to apply an online questionnaire to the participating youngsters where we inquired about their knowledge (from knowing nothing to knowing very well) about digital tools, including those of a multipurpose and communicational nature and those more specialised for mathematical purposes. Besides, we were able to complement and connect the results from the online questionnaire with other qualitative data originating from interviews with participants and former participants and also with some of the teachers who have followed the competitions over time, and mostly with samples of digital solutions collected over the 3 years lifespan of the Problem@Web project. We now highlight some of the most significant results obtained from the analysis of those data.

Our respondents' sample corresponds to a return rate of about 20 % of the youngsters who were participating in the competitions that lived and attended school in the Algarve region. We have obtained a balanced distribution in terms of gender, but on the school level the 5th grade was predominant, followed by the 6th grade. Also on age, most of the respondents were 11 years old followed by the 12-year-olds. This reflects the prevalence of younger students in the competitions, especially in the 5th and 6th grades, which seem to be the more enthusiastic and those who more easily adhere to these challenges, possibly because of a lower load of school subjects or due to a more playful experience with mathematics and a lesser fear of failure when compared to the older students in 8th grade.

The questionnaire showed that computer availability, Internet access and access to an e-mail account were virtually universal in the homes of participants. In addition, almost all respondents considered that they knew well or very well how to use the Internet and e-mail, revealing that the youngsters were perfectly able to use online communication in order to participate in the competitions, something that was not completely true early in 2005. At that time, many parents and teachers gave support and served as transmission belts between the youngsters and the organisation of the competitions.

Some interview data also showed that computer use has been for many of them an attractive factor that motivated them to participate in the online competitions. Some of the youngsters actually reported that over the years of their participation in the competitions, they had learnt to take advantage of the potential of several digital tools that initially they had only slight knowledge. It is noteworthy that one of the interviewed youngsters, with visual impairments, was confident in stating that the use of the computer in the competitions helped him to overcome difficulties with writing and reading and gave him the will and the tools to get involved with and enjoy solving mathematical problems.

With regard to analysing the skills in using technological tools, we organised our analysis from the use of written text to solve and express the solutions to the problems, through the use of diagrams, tables and images, to the use of features of more specific software directly linked to mathematical procedures such as algebraic relations and geometric constructions.

As we have already noted, there remained participants who preferred to use paper and pencil to prepare their solutions, with the solutions being digitalised afterwards using a digital camera or a scanner. Nevertheless, the vast majority of respondents said that they knew about text editing with Word and about 70 % knew about writing text in Excel.

Over time, various solutions were presented by the participants in a clear, consistent and effective way; in essence, these were quite detailed descriptions of the solving process and strategy presented through natural language. The use of colours, icons or arrows to highlight specific information or results was also common as well as other ways of using editing add-ins to make a more interesting or effective explanation (see the solution presented in Fig. 8.3 where the "heart" operation between numbers had to be discovered based on some given cases).

We also sought to know the capability that youngsters felt they had in the construction of tables, both in Word and in Excel. In fact, the use of tables was quite often chosen by many participants in several of the proposed problems. In many cases, such tables were created with the text editor or with the spreadsheet or even with the online form in the website. We found that youngsters considered themselves better able to work with tables in Word than in Excel. Indeed, the knowledge of Excel seemed relatively low, and this became more accentuated as the functionality of using Excel in computational aspects such as the use of formulas or the creation of graphs was considered. An interesting aspect that emerged from many of the resolutions presented in Excel was that youngsters used the grid to create quite varied ways of organising, schematising and analysing the elements contained in

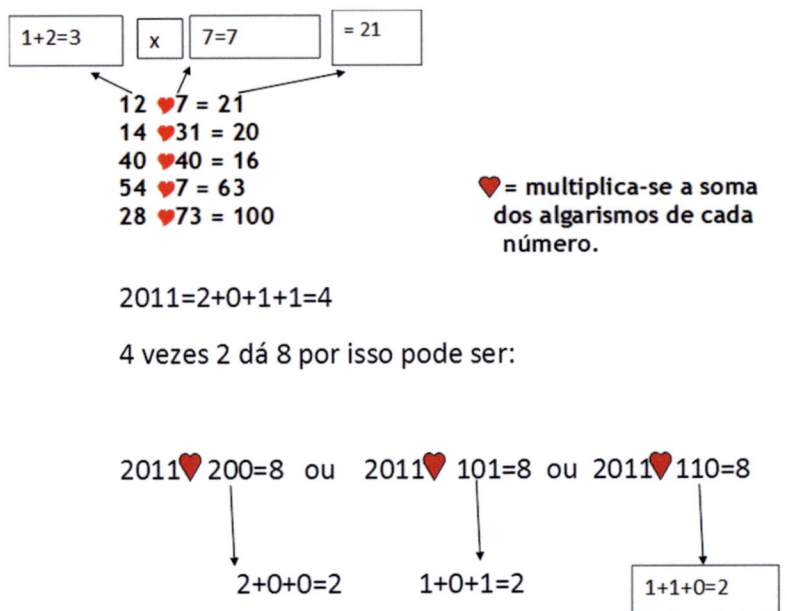

Fig. 8.3 Solution sent as a Word file submitted to Problem #4 of SUB12 (2010–2011 edition)

various problems. In many cases, they filled cells with colour, inserted images or created visual representations that revealed patterns, for example. Another indicator that we obtained through the interviews was that youngsters seemed to learn to use Excel more at home than at school and it was often parents who helped and encouraged learning of the use of the spreadsheet.

The use of images, pictures and diagrams was also an element very much present in the youngsters' answers to the problems, and this revealed much of their representational capability in solving and expressing. They used images that they sought out to illustrate their ideas and their reasoning. They did this with Word, with Paint or with PowerPoint, among others. When answering the questionnaire, a large percentage of the participants claimed to have a very good knowledge of the use of images with Word, Paint and PowerPoint. In short, the category of using images and diagrams was heavily rooted in a visual and pictorial form of solving and expressing and seemed to persist in all the editions of the competitions and across all ages of participants (see Fig. 8.4 for an example of a solution in Word where the pictorial form of exposing the solution is significant and speaks for itself).

In contrast, the use of numerical software was less common than the previous categories of tool use, but participants who employed Excel used it for constructing numerical relations with formulas and took advantage of other computational and mathematical features like the construction of graphs (e.g. pie charts).

Reliance on geometry software appeared much less in solution examples. GeoGebra was unknown to almost 60 % of respondents. Participants who opted for GeoGebra used it mainly to solve geometry problems, usually for the construction of figures, for

Fig. 8.4 Solution sent as a Word file to Problem #7 of SUB12 (2010–2011 edition)

creating diagrams, measuring or plotting a function. Some of the interviewed participants admitted to having had their initial contact with GeoGebra at school, but they said to have developed their knowledge of the software independently at home.

Through these different types of data, it can be seen from the profile of the youngsters who participated in the online competitions that digital technologies played an important role both in their solving of the mathematical problems and in their ways of expressing and communicating their mathematical thinking as it developed during that process.

We can conclude that the youngsters' capacity to put forward their mathematical thinking through natural language, through tables and various forms of visual representation (pictures, colours, diagrams, images, icons, etc.), is strongly linked to their problem-solving capability and thus their digital discourse is a digital-mathematical discourse that reflects the affordances that young people spontaneously identify and effectively use within the competitions. We must not fail to notice that much of their digital fluency is still far from the more technical and mathematical tools aimed to deal with numerical or graphical processes. Moreover, it is important to retain that respondents seemed to learn more about specific mathematics-related technological tools at home than at school. The SUB12 and SUB14 competitions are clearly an opportunity for the youngsters to disclose their knowledge, develop it and put it into practice in solving and expressing moderate mathematical challenges.

In the next section, we illustrate the perspectives of teachers on youngsters solving mathematical problems with technology during the SUB12 and SUB14 mathematics competitions. Drawing on a series of interviews with teachers who have supported the participation of their students over several editions of the competitions, we focus on what they see as the competitions' most significant features.

8.4 The Perspectives of the Youngsters' Teachers

Various interviews with teachers who accompanied their students during the competitions generated a set of relevant information that allows us to characterise: (1) their role among the youngsters in engaging in this context, (2) the main aspects of the competitions that they consider pedagogically important for the development of competencies in their students and (3) what they draw as the more meaningful experiences from supporting and encouraging their students.

We can conclude that the involvement of teachers varied, often depending on the momentum of their schools and their peer groups and on the value they attach to the participation of youngsters in beyond-school projects and to the importance of attracting students to mathematics by stimulating their abilities and enriching their learning experience. In general, all the teachers accepted with pleasure and diligence the task of informing their students about the SUB12 and SUB14 competitions, motivating them, giving information, distributing posters and flyers and referring the benefits that students can get by engaging in mathematical problem-solving work (Fig. 8.5).

The level of involvement of a certain cluster of teachers went far beyond this mission of encouragement and information. These are the teachers who systematically followed their students in their participation, providing support, discussing the problems and giving clues and help at various times. Many of these teachers used the problems of the competition in their classes (whether in mathematics classes or in supervised study lessons), leaving students the decision to engage or not in the contest. This illustrates how the teachers considered that many of the proposed problems were relevant and useful for work in the classroom. Finally, there are teachers who were even more committed through the various phases of the competitions and who got in touch with the organising team, accompanied their students to the Final and offered their collaboration (see Fig. 8.6 of a teacher together with one of her students at the Final of SUB14). This is a fact of great importance that leads us to recognise the commitment and interest of teachers as one of the major factors for the success of these competitions. Given, as we said earlier, that these are inclusive competitions that seek to include students of different attainment levels, the statements of the teachers help to confirm that the SUB12 and SUB14 competitions count as enrichment initiatives that work in a friendly and supportive way and are close to schools and families offering an opportunity to engage youngsters in moderate mathematical challenges.

Teachers clearly indicated that one of the elements they valued most in the competitions was indeed the problems proposed. They found them suitable for their students with a degree of difficulty that they see as fitting, interesting to work in the classroom, different from the typical school problems in textbooks and allowing a wide variety of approaches and strategies to address them.

The interviewed teachers described how they used the problems of the competition in their classes. Some said they could be useful either for introducing a certain mathematical topic or for practising a given content, or could serve as a means to

Fig. 8.5 Image of a poster announcing the competitions SUB12 and SUB14 that was distributed to all the schools covered by the competitions

Fig. 8.6 Photograph taken at the awarding ceremony of a teacher with one of her students who was ranked first at the Final of Sub14

develop mathematical reasoning and the construction of mathematical connections between various concepts. They mentioned that the problems were a pedagogical resource for their teaching practice and particularly liked to solve them, often as soon as they were posted on the website. For several of the teachers, the word that best describes the nature of the problems is the word challenge. They felt challenged and believed the problems to be stimulating for the youngsters too. They liked to think of several ways to solve the problems before they discussed them with their students but admitted they were often surprised by the unexpected and original strategies and ideas of their students, quite different from those that are associated with standard approaches based on curricular content. In this sense, they acknowledged in their students not only the capability to deal with the problems but also their mathematical creativity that allowed them to find relevant ways to achieve a solution and to think mathematically regardless of whether they already learnt this or that content.

The support from teachers to their students was marked by the willingness to help students to improve their problem-solving capability and instigate them to think and try to reach a solution on their own. The interviewed teachers were unanimous in saying that they did not show the students how to get the correct answer. Rather, they promoted discussion among the students in class, provided clues, instructed them to think more at home. Subsequently in the classroom, they said that

8.4 The Perspectives of the Youngsters' Teachers

they compared the strategies of several students or used the problems to move forward to other mathematical questions. Some also reported that they made themselves available to offer any assistance to any student who needed this both in class and outside the class or even by e-mail. In many cases, they checked if the problem was correctly solved before the student submitted their answer to the competition. While responding to the possibility of participants seeking help that is clearly advocated by the SUB12 and SUB14 competitions, teachers interpreted the problem-solving activity as a learning situation that required the engagement of youngsters and was powered by exchanging ideas, using heuristics and the capacity to think mathematically. Therefore, they placed the autonomous work of the student as a primacy of their teaching of problem-solving.

Highly valued by the teachers was also the importance of mathematical communication and expression during the competitions. Solving and expressing the whole process that leads to the solution of a problem was a demanding task for youngsters, and teachers were aware of that fact. For teachers themselves, it was a challenge to help students and give them tools and tips on the activity of presenting the resolution in a clear, explicit and mathematically correct way. For some interviewees, there were students for whom learning mathematics was synonymous with explaining things briefly and condensing everything in calculations or algorithms. Leading students to present their ideas, to expose their reasoning and to use forms of presenting their strategies was a challenge for the teachers who started to invest more in the development of mathematical communication in their practices. For example, some teachers began to explore further the use of diagrams and other representations or the use of natural language to describe the solving processes in detail. They stressed however that there are several possible ways to present the resolution of a problem and, accordingly, shared the aim of giving participants in the competitions freedom to decide on how to solve and express a problem and on the tools they deemed most effective for that end.

The teachers saw the centrality of the expression of mathematical thinking as a distinctive feature of these competitions. Along with the type of problems proposed, which are considered challenging but not selective, they found that the SUB12 and SU14 competitions were closer to the students and to the nurturing of their skills than other competitions to which they also attribute value.

The other point that is highlighted by the teachers was undoubtedly the usage of technology and the Internet as a means of connecting the participants to the competitions. Some stated the difficulties they faced at the beginning, almost a decade ago, with technical details such as Internet access and the creation of e-mail accounts for their students. They genuinely reported what they learnt from their students over the years and referred to parental help as having been equally important to the participants on this matter. They also believed that youngsters were able to handle ICTs very expeditiously to address and express their solutions to the problems.

It is important to note that some of these teachers felt unprepared for a fruitful use of technology with their students, but this did not stop them from encouraging students and sharing with them the possibilities offered by technologies that have become more available in schools over the course of time. It is reasonable to conclude that these teachers faced, without any resistance, the use of technology by their students and that they actually collaborated in this perspective. In fact there were several times when the resolutions submitted by the participants even explained how the digital tool worked, as in the example of Fig. 8.7 where two students used Excel and explained how the formulas should be introduced and asked the SUB14 organising team to try this out.

Like the organising team of the competitions, the teachers appreciated the suggestions that came from the students, and in other cases they themselves gave a push for the use of technologies. Clearly, the teachers realised that the use of technology was very much involved in the SUB12 and SUB14 competitions and that digital technologies were put to use for solving problems and concomitantly expressing the underlying processes.

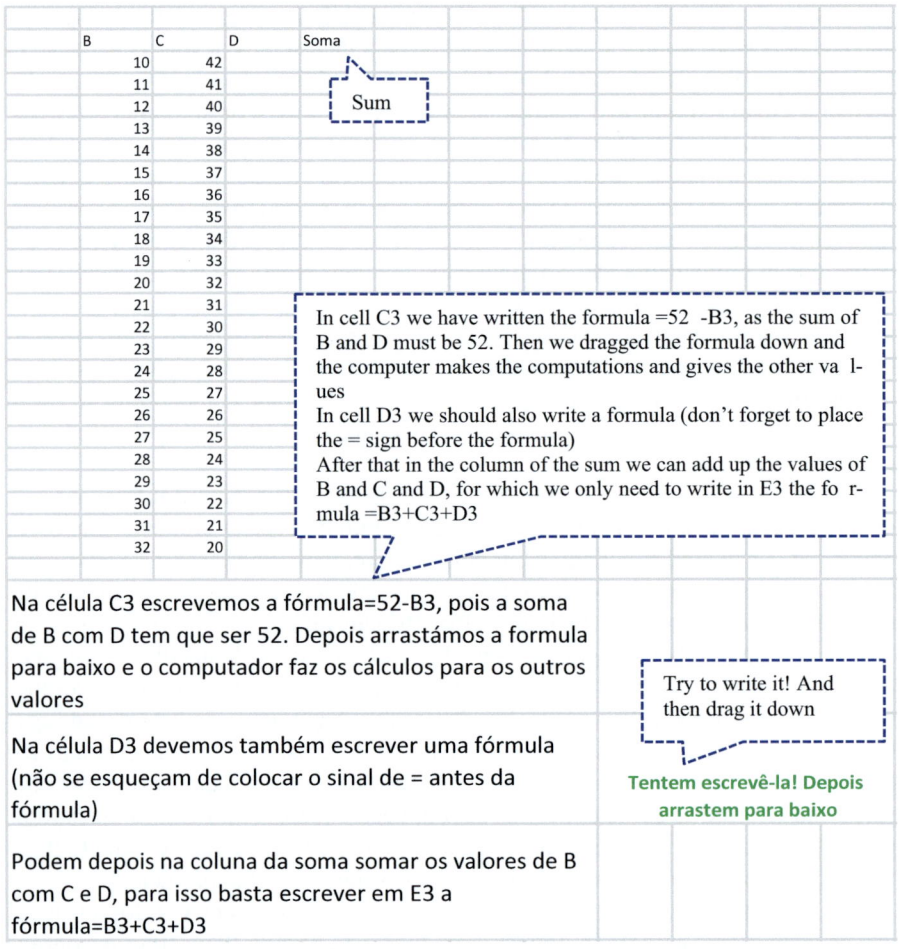

Fig. 8.7 A solution obtained with Excel where the students who produced it explain how it was done and challenge the receivers to try it

In this section, we recognise the work and involvement of the participating teachers, through their contribution in the different stages (dissemination, qualifying and final). We collected evidence that the teachers played a decisive role in the success of the online competitions. Not only did they promote the participation of their students in the competitions, they also strongly valued the mathematical problems posed during the various rounds of the competitions.

In the next section, we provide the theoretical perspectives that framed the Problem@Web project's study of a specific and relatively new phenomenon, that of youngsters solving mathematical problems with the digital technologies of their choice.

8.5 Theoretical Framework

The background of the study is centred on a particular theoretical stance from which we consider the problem-solving process as a synchronous process of mathematisation and of expressing mathematical thinking; such a perspective is additionally supported by two specific research backings: (a) the role of external representations in problem-solving and expressing and (b) the symbiotic relation between the individual and the digital tools in a problem-solving and expressing technological context (see Fig. 8.8 for a diagram of the fundamental theoretical components).

The rationale from which derives this theoretical framework lies in the research setting and in the type of data that we planned to analyse. We wanted to know more about the way the young participants in these online competitions find effective and productive ways of thinking about various problems and communicate mathematically their solution processes based on the digital resources that they choose for some reason. We saw above in the youngsters' answers to a questionnaire, in inter-

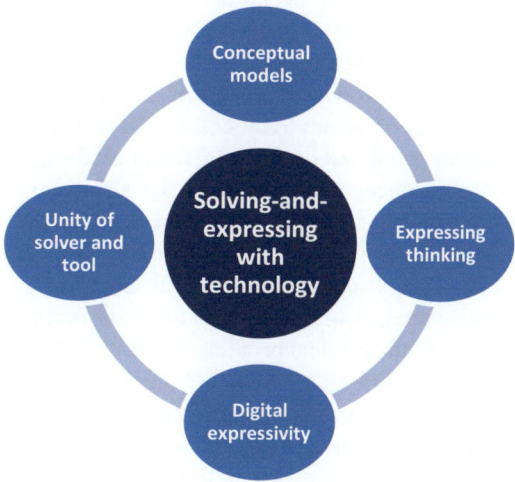

Fig. 8.8 Solving and expressing with technology as a core concept enclosed by theoretical views on problem-solving activity and its relation with the use of digital tools

views with participants and teachers as well as through the solutions submitted during the qualifying phases of the competitions that there was a great familiarity of youngsters with some digital tools of widespread use. Nevertheless, we also saw that for other more specific tools and those more aimed at mathematical processes, the fluency of the youngsters was less. In any case, it stands out that there is an important role of technology in the development and exposure of the solutions to the problems, whatever tool may be used.

Leaving out those solutions that participants produced with paper and pencil and submitted as photos or scanned images, we are left with a variety of ways of explaining different approaches to the proposed problems that are also expressed with varying digital tools. We have, so to speak, multiple narratives produced by youngsters who illustrate and display their way of thinking embodied in multimedia representations and discourses. We consider that these narratives offer a digital-mathematical discourse that has several distinctive marks, including the capability to use the representational expressiveness of digital technologies and a strong link between how to think mathematically and the use of the chosen technology. This allows us to envisage solving a mathematical problem as a synchronous process of mathematisation and expression of mathematical thinking in which digital tools play a key role.

As extensively illustrated in the earlier chapters of this book, the proposed problems in these competitions are characterised by having a moderately challenging nature that gives the possibility of various forms of resolution. Each of these different forms results from viewing the situation from a mathematical perspective and unravelling a conceptualisation of that situation that includes understanding and representing it. We therefore focus on the mathematisation of problem situations and draw on the development of problem-driven conceptual models (English & Sriraman, 2010; Gravemeijer, 2002, 2007; Lesh, 1981, 2000; Lesh & Doerr, 2003a, 2003b; Lesh & Zawojewski, 2007).

A conceptual model of the situation includes explicit descriptive or explanatory systems. It is that descriptive and explanatory quality that makes it function as a model, i.e. an externalisation of the ways in which individuals are actually interpreting a situation, even if the solver may not draw on formal, school subject knowledge. The use of pictures created with drawing digital tools, or the construction of tables with text editors or spreadsheets, as well as the use of dynamic calculations with Excel or geometrical constructions made with dynamic geometry systems are a fundamental part of the representational expressivity of youngsters' narratives and of their conceptual models of the situations. Thus we argue that strategies and representations are variously combined by different youngsters in digital environments for problem-solving and expressing. This clearly disrupts the idea of "optimal" strategy and "optimal" representation to approach a particular problem and rather increases the awareness that different tools offer different ways to solve and explore mathematical problems (Barrera-Mora & Reyes-Rodríguez, 2013; Hegedus & Moreno-Armella, 2009a, 2009b; Santos-Trigo, 2004, 2007).

The context of youngsters' independent use of the computer, having as essential elements the choice of the tool and the youngsters' recognition of affordances in the tool tuned with their ways of thinking in search of a solution, also requires addressing the relationship between technological media and cognitive activity. Following the theoretical perspectives of humans-with-media (Borba & Villarreal, 2005) and co-action between the individual and the digital tool (Moreno-Armella & Hegedus, 2009; Moreno-Armella, Hegedus, & Kaput, 2008), we endorse the vision described by Moreno-Armella and Sriraman (2010) on the computer's dual role in youngster's cognitive activity: a mirror which reflects actions and an agent who acts:

> The computer will be transformed, gradually, into a cognitive mirror and a cognitive agent from which students' learning can take considerable profit. It is not anymore the simplistic idea "the computer is doing the task of the student" but something new and radical: providing students with a cognitive partner. (p. 224)

The indivisible unity of the solver and the tool is strongly supported by the research on the use of interactive tools for learning mathematics, but it can be reinforced and further justified by the new possibilities of communication and expression provided by digital technologies—imagery, visualisation, manipulating objects on the screen, formatting, editing, using colour and highlights, organising information and structuring schemes involving symbols, pictures or sounds are examples of a host of possibilities. The thinking and the communication of the thinking are becoming more and more blended under the increasing flexibility of digital tools.

The inter-shaping relationship between the user and the digital tool in expressing mathematical thinking, i.e. the way of transmitting a conceptual model, is an activity in which the intentionality interacts with the executability of digital representations. Digital solutions submitted by participants in the competitions are therefore more than a transposition of a conceptual model to a digital medium. The use of digital tools, either the simpler and more earthly or the more sophisticated and robust, is more than a vehicle; it represents the co-action between the subject and the tool in the construction of conceptual models and concomitant expression of mathematical thinking in solving problems.

Based on this theoretical view, we use the framework synthesised in the previous figure (Fig. 8.8) as a kind of a toolbox to illuminate the analysis of some selected problems. In each case, due to varying nature of the problems, we integrated an additional theoretical background that provides ideas and more specific concepts, judged useful for a detailed analysis of the data obtained.

In the following section, we report the analysis of three problems, widely varied in terms of the mathematics involved: a problem of invariance in geometry, a problem of quantity variation in algebra and a problem of co-variation in a relative motion situation.

8.6 Digitally Expressing Mathematical Problem-Solving

This section summarises our analysis of how the youngsters expressed and communicated their mathematical problem-solving through their use of digital tools of their choice. Using the theoretical frame outlined above (and captured by Fig. 8.8) and using both data-driven encoding and theory-driven coding (as noted further above), we conducted a set of three analyses that involved, in turn, the mathematics of invariance, variation and co-variation. Full details of these analyses are presented in the previous three chapters of this book. Here, we condense the main points.

8.6.1 Digitally Expressing Conceptual Models of Geometrical Invariance

Drawing on the theoretical ideas above, we add some additional arguments relevant to analysing data concerning the usage of GeoGebra for solving geometry problems posed during SUB14. We focus on the perception of the affordances of a digital tool as a crucial element for its effective use, i.e. the perception of how a particular tool will help in solving the problem and expressing its solution; we also argue that the perception of affordances emerges from the interactions between the solving agent and the tool, which means that the tool's affordances and the individual's aptitude cannot be determined in the absence of one of them. This assumption has implications for the way of looking into the data collected in that it requires noticing the types of GeoGebra's utilities that youngsters were using to solve a problem of invariance. Consequently, a set of categorised affordances that were repeatedly used by youngsters who submitted answers with GeoGebra was built as a starting point for describing the type of underlying approach. This approach was then further analysed in light of the kind of robustness of the conceptual model underpinning the solution achieved.

Drawing on the metaphor "humans-with-media" that encapsulates the idea of the indivisibility between the solvers and the tools with which they solve problems, we also aimed to show how the technological tools used to communicate and produce or represent mathematical ideas influence the type of mathematical thinking and mathematical representations that result from this activity. Different groups of humans-with-media originate different ways of thinking and knowing. In fact, participants who submitted resolutions produced in GeoGebra revealed different approaches to the same problem, and our goal was to differentiate these approaches despite the fact that they all used the same technological tool. A step forward in this direction was provided by analysing the mathematisation activity in terms of the shift from an informal model (where the solver focuses on the relation between the context and the mathematical concepts or procedures) to the development of a more formal and general model (detached from the context) focusing on symbolisation and abstraction.

We analysed four solutions to the problem "Building a Flowerbed", which encompasses exploring the invariance of the area of a changeable triangle, by modifying the geometrical figure. The data was selected from a diversity of answers offering evidence of the use of GeoGebra.

We started by *zooming in* on the participants' productions and, by using the construction protocols whenever possible, we traced back every step of the construction thus allowing us to identify the affordances perceived in each case to construct the figure given in the problem situation and produce a conceptual model of the situation to obtain the required answer. Then we made a *zoom out* over the four solutions by comparing and contrasting them according to the levels of mathematisation involved and the consequent robustness of the underlying conceptual models. In what follows, we present a summary of the main results of our analysis. Here, all solvers perceived the convenience of GeoGebra to represent the situation described, foreseeing the transformation of the static images in the statement into dynamic figures and ideas (e.g. the changeable triangle associated with moving the position of the stick).

Solving-with-GeoGebra was clearly exhibited in the analysed answers in that the solution and its presentation, justification or explanation come out through the symbiosis between the affordances of GeoGebra and the youngster's aptitude in dealing both with the tool and the aim of the problem. Thus, the constructions, the strategies, the results and justifications presented in each case sum up a conceptual model of the invariance of the area and expose the activity of these "youngsters-with-GeoGebra". The digital tool, GeoGebra, was used differently in each of the cases: (1) to construct and measure, (2) to conjecture, (3) to verify and (4) to think-with. This shows how the digital tool influences the type of mathematical thinking produced which, in turn, influences the conceptual model that is developed. The conceptual models developed by these youngsters-with-GeoGebra ranged from a horizontal mathematisation (attached to the context and to the obvious confirmation provided by the tool) to a form of vertical mathematisation (where the youngster creates a mathematical proof that validates the model).

These findings illustrate the ways in which GeoGebra affords different approaches to one problem in terms of the conceptual models developed for studying and justifying the invariance of the area of a changeable triangle. We interpret the different ways of dealing with the tool and with the mathematical knowledge as instances of youngsters-with-media engaged in a "solving-with-dynamic-geometry-software" activity, enclosing a range of procedures brought forth by the symbioses between the affordances of the dynamic geometry software and the youngsters' aptitudes. The analysis showed that different youngsters solving the same problem with the same digital media and recognising a relatively similar set of affordances of the tool produce different digital solutions, but they also generate qualitatively different conceptual models, in this case, for the invariance of the area. This is solid evidence of how the spontaneous use of technology changes and reshapes mathematical problem-solving.

Below, we pursue the same line of analysis by trying to identify how youngsters within the classroom and within the SUB14 solve algebraic problems of quantity variation when they decide to make use of the electronic spreadsheet.

8.6.2 *Digitally Expressing Algebraic Thinking in Quantity Variation*

The two problems we have selected from SUB14 regarding variation and varying quantities involve, from the formal algebra point of view, solving inequalities and systems of simultaneous equations. The youngsters participating in SUB14 lacked this formal algebraic knowledge and had to tackle them using other means and forms of mathematisation; nevertheless, some of them have chosen to use the spreadsheet to devise a way to get the solution.

In studying the youngsters' approaches, we first looked at aspects which related to the use of an algebraic-like syntax elicited by the use of spreadsheet columns and formulas. In particular, we aimed to understand how the use of this digital resource was reflected in representations that maintained a strong connection to the algebraic thinking of the participants. Therefore, we decided to translate systematically the relations expressed by the students in the spreadsheet (by using formulas, automatic generation of sequences, variable-columns or variable cells) to an algebraic symbolic language. This provided us with a clear way to appreciate the conceptual model (a productive way of thinking mathematically about each of the problematic situations) underlying the search for the solution.

We complemented the theoretical framework outlined above with a focus on algebraic thinking and its essence, especially stressing that the use of algebraic symbolism does not necessarily capture the capability to think algebraically. This means, for example, that the absence of symbolic algebraic language does not prevent understanding the role of variables (independent and dependent) or the establishment of relations between them, often involving several composite relationships.

One relevant concept that we further developed was the concept of co-action, based on which we have focused on the way that students are at the same time guided by and guiding the interactive environment of the spreadsheet. In particular, we observed how the participants defined their problem variables, how those were organised in the table, how they set dependencies and how they inspected the numerical outputs, interpreting them in light of the restrictions given in the problems. We therefore observed that the solution to each of the problems emerged from a student's partnership with the tool. It allowed them simultaneously developing productive ways of thinking and expressing such thinking on variation through the digital media, while recognising and using the affordances of the tool.

Besides considering the data provided by all the solutions submitted by the students who were engaged in the online competition, the same two problems were also given to a class of 8th graders as part of their periods of Supervised Mathematics Study (i.e. non-curricular classes) in a public middle-school located in the south of Portugal. In the classroom, we had the opportunity to record the voices of the students and their actions in the computer with screen-recording software while they were solving the problems, some in pairs and others individually as they preferred.

In the case of the first problem, it was evident a certain progression in the types of resolutions found, taking into account a criterion which quickly became evident: the increasing use of variable-columns instead of the use of recipient cells just aimed to introduce number inputs with which operations were carried out. This allowed us to depict a certain progression or a kind of hierarchy of the solutions analysed and the respective conceptual models of the situation. However, in the second problem, this criterion proved to be ineffective because the variations that occurred, although many, did not rely on this feature of the use of the spreadsheet.

Other ways to distinguish the different strategies and ways of organising the inputs and outputs in the spreadsheet have to be considered, and this is an open field that future research should continue. As a feasible possibility, we simply suggest that the greater or lesser number of intermediate relationships between variables created by the solvers—something that has become highly visible in various resolutions, sometimes with the use of two independent tables that were related at the end with the goal of finding common values—may be a promising avenue. Undoubtedly, we can conclude that the various proposed solutions have shown the importance of students' ability in identifying all the variables (and constants) and all the conditions connecting them and the restrictions involved.

The results strongly support the idea that the spreadsheet was, for the youngsters who chose to use it, a suitable and productive environment for the development of their conceptual models; each of the problems on quantity variation and the conditions that shaped it were understood in very different ways and led to the concatenation of relevant functional relationships describing the variation of quantities subject to restrictions. In summary, the spreadsheet offered a dynamic environment for the exploration and solution of such problems (quantity variation problems) and for the expression of students' algebraic thinking, namely, through the spreadsheet-based representations.

It was apparent (namely, from the data collected in the classroom) that the development of different conceptual models is related with the co-action between the user and the tool. We have seen that the spreadsheet provided a means of experimentation, of trial and refinement, a context that although involving variables expresses data numerically, making the results of experimentation easier and more immediate to evaluate and therefore more inducible to reaction from the student.

The data from the classroom were important to understand in more detail the ways students approached the problems, their attempts, discussions and the co-action that takes place in the process of problem-solving and also to see how the expression of the algebraic thinking emerged associated with each production.

Different equations and dependency relationships between variables can translate the same problem and that means an important step for understanding algebraic transformations. Therefore, the level of fluency of the solvers who used the spreadsheet on the two problems represents an indicator that this digital tool has great potential within the youngsters' modes of reasoning, despite an overall lack of familiarity with this tool that youngsters have reported as revealed by the results of our research (see Chap. 2).

The spontaneous choice and use of the spreadsheet revealed by the youngsters whose solutions we have analysed have shown that their productions with the spreadsheet can relate with formal algebra and may well support the learning of algebra.

We next turn to a problem involving co-variation and of the common-usage digital tools that youngsters used for translating their conceptual models of relative motion, showing that such models are quite distinct from the formal algebraic models used by mathematicians.

8.6.3 *Digitally Expressing Co-variation in a Motion Problem*

In concentrating our attention on a problem involving motion (i.e. requiring relating the variables space and time), we had to expand our theoretical toolkit in order to integrate perspectives on co-variational reasoning and the modelling of dynamic situations by means of suitable representational modes.

Co-variational reasoning has to do primarily with the coordinating of the simultaneous changes of several variables. Research has shown that youngsters face difficulties in this type of coordination and in the modelling of dynamic situations and it is also something that teachers define as being complex (Carlson, 2002; Carlson, Jacobs, Coe, Larsen, & Hsu, 2002; Carlson, Larsen & Lesh, 2003; Zeytun, Çetinkaya, & Erbaş, 2010). One aspect to consider is that co-variational reasoning is strongly related to the development of kinaesthetic images of problem situations, a fact that reflects the importance of building mental images of motion, engaging with metaphorical thinking, physical enactment and bodily references. This seems to be of importance in youngsters' identification of the relevant variables and in their sense-making of the effect of changing one variable over another one. We also began a discussion about visualisation insofar as this capacity is closely associated with the construction of effective representations to express the coordination of two variables changing in tandem. In fact, some studies have shown that youngsters have a great capacity to invent their own ways of representing and translating motion situations into visual forms of representation.

The aim of our analysis was to understand the conceptual models and types of mathematical representations that youngsters used to solve and express a motion problem and how their models were related to the digital tools to which they resorted. The motion problem was launched early in the SUB14 competition, being the first of the problems proposed in the 2011–2012 edition. Given the large volume of answers received from all participants involved in the competition, we chose to focus only on the answers received from the participants attending 8th grade. First, an analysis of how the problem was solved by a group of experts (mathematicians) was undertaken. Next, a categorisation of the solution strategies used by the experts was developed. Finally, the categorisation was used to sort and classify the 8th graders' solutions. In examining the youngsters' solutions we aimed at identifying the youngsters' prevalent conceptual models and at relating them with the technological tools they employed in their solving and expressing of co-variation.

We found that the dominant conceptual model in the resolutions of youngsters (that we called model 2) may be described as "the developing journey", which means looking into the future and reconstructing the motion step by step as time goes on. In contrast, experts have tended to use model 1 that can be described as "the completed journey" wherein one imagines the journey already completed and looks back seeking to determine the elapsed time.

Within model 2, the forms of representation that were identified in youngsters' answers to the problems were tabular representations, diagrammatic representations, verbal representations and pictorial representations, with the verbal representations being the most frequent. The third more often used representation form was the diagrammatic one, with a percentage very close to the amount of tabular representations. It represents a strong contrast with the rare presence of diagrams in the experts' solutions, and it showed how youngsters sought to show information visually organised and denoting the idea of a temporal sequence of the relative positions of the two moving bodies. The fourth type of representation was one that was not revealed at all in the experts' solutions, and it refers to pictorial displays of movement involving pictures, the use of colours and captions in a way that aims to capture the dynamic characteristic of the situation thus providing a quasi-dynamical expression of the solution to the problem.

Using technology as opposed to formal algebraic language for solving and expressing co-variational reasoning was clearly the option of the large majority of youngsters and proved very effective in youngsters' approaches. The type of representations used by the youngsters highlight the visual aspect as inherent and fundamental in the problem-solving process (quite noticeable in schematic approaches where the reconstruction of the motion is performed). The schematic/diagrammatic representations and the pictorial/figurative ones are static representations but allow depicting a dynamic situation, thus revealing the youngsters' capability to coordinate the co-variation of time and space. These seem to be indicators of a special kind of visuality, consistent with the multimedia language of everyday digital technologies that youngsters apparently are comfortable with and accustomed to use. This digital visuality is a very clear attribute of the pictorial/figurative representations observed in a number of solutions. In any case, throughout the several categories of representations pertaining to model 2, it is possible to envisage how, in the case of solving a motion problem, young participants in the competition were performing as humans-with-media. Our results indicate that youngsters lay hold of the tools that are more easily available to them, which in many cases simply means the use of a word processor or of the e-mail window. There were also many participants who proved to be fluent in the construction and formatting of tables or in presenting data in tabular form with common tools. Finally, the nature of many graphical representations reveals a digital enactment of co-variation as if the situation were being described by way of frames. This kind of visual, graphical and pictorial expression was supported by the efficient use of multimedia documents created from word processors, PowerPoint slides and other software that allows editing of drawing objects.

In the following final sections of this chapter, we outline a global view of the results of the project, seeking to highlight the most relevant conclusions from the broader perspective on the experiences reported by youngsters and teachers in solving problems with technology over the competitions, to the tighter focus on the approaches generated by participants in certain mathematical problems with digital technologies (GeoGebra, Excel and also multipurpose digital tools). We then look to the implications of the findings and suggestions for further research.

8.7 Discussion of the Findings

In our Problem@Web project, our aim was to produce a global picture of the young participants in the mathematical competitions SUB12 and SUB14 regarding their knowledge and reported usage of technologies as a background to investigating what they were capable of achieving when using digital technologies in problem-solving and in expressing their solutions mathematically. Notwithstanding the many reports already published around the world on the digital literacy of twenty-first-century youngsters (including Biagi & Loi, 2013; Fraillon, Ainley, Schulz, Friedman, & Gebhardt, 2014; Kadijevich, 2015), it seems that an accurate perception of this digital generation is still not fully achieved. In particular, there seems to be indicators that these youngsters surrounded by digital technologies reveal some weak skills in using technology as a means for learning, tending to be more resourceful in uses related to social communication and retrieving online content for fun and recreation.

Our data showed to some extent that this is also the reality of the youngsters we have questioned. They have domain over a set of general-use digital tools, but it appears that they are less aware of the digital resources with a stronger association with mathematical knowledge and procedures. Meanwhile, we also realised that the young participants in these competitions discovered many of the capabilities of their computers by getting involved in solving problems and seeking expeditious, appropriate and productive ways of expressing their mathematical thinking. In fact, although their technological skills were mostly directed to other activities, they were able to harness them for the production of their problem resolutions and were simultaneously developing and improving their capacity to create and use several mathematical representations (pictorial, schematic, tabular, textual and so on).

The large number of problem resolutions that we analysed over the Problem@Web project became therefore a clear revelation and provided evidence of what youngsters can do in their own time and with their spontaneous choices of digital tools. An important aspect we want to highlight is the distinction between these digital productions and those that are more common in the mathematics classroom using paper and pencil. In many cases, these digital productions reflect a strong interaction between the tool and the way the youngster approached the problem to arrive at the solution and communicate it to the organising team. In addition to depicting the result of this interaction as a digital-mathematical discourse, we were also able to explain it as a result of the co-action between the tool and the solver (Hegedus, Donald, & Moreno-Armella, 2007), thereby illustrating this reciprocity as an instance of the humans-with-media unity (Borba & Villarreal, 2005). This interconnectedness between the technology and the mathematics involved in a problem leads to jointly developed technological skills and mathematical skills, resulting in the capacity of mathematical problem-solving with technology, as we depict in Fig. 8.9.

The in-depth analysis we have performed with three very different kinds of problems—invariance, quantity variation and co-variation—has allowed us to

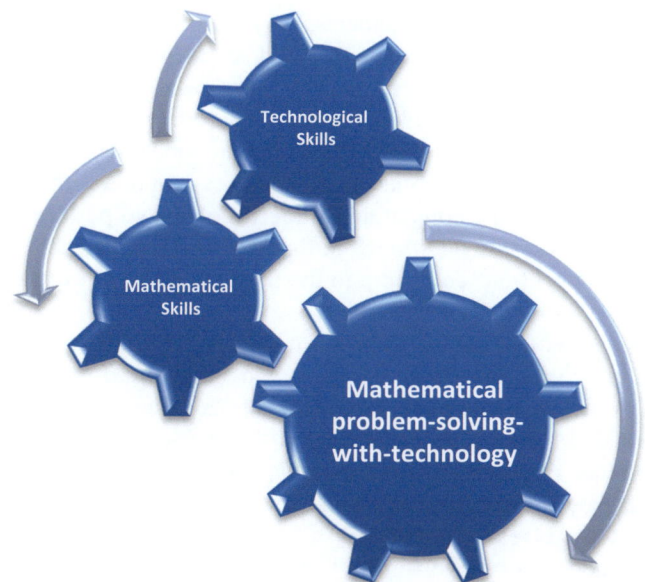

Fig. 8.9 Mathematical problem-solving with technology as an output of the joint development of technological skills and mathematical skills

understand more clearly what is at stake in solving mathematical problems with technology. We found that there is a strong cohesion between the conceptual models that youngsters develop to find the solution of a problem and the affordances they perceive in the tools they have chosen in each case.

Different uses of GeoGebra showed different ways of thinking and expressing mathematical ideas on the invariance of the area of a triangle; also different uses of Excel have provided different models that led to the solution of problems with an unknown under various restrictions; finally, the resolution of a motion problem demonstrated the effectiveness of the use of several multipurpose digital tools to express a predominant conceptual model of time and displacement changing in tandem.

The results of our analyses also show the centrality of the representational power of digital technologies that youngsters clearly dominate and exploit. The representational expressivity emerges from the productions of the youngsters when we identify the concomitant and inextricable processes of solving and expressing in mathematical problem-solving (rather than looking at those as isolated processes). The solving and expressing in youngsters' productions is therefore largely associated with their spontaneous uses of technologies and became a distinguishing quality of many of the creative solutions they have presented.

One of the aspects seen as crucial by the teachers we interviewed was the nature of the problems posed in the competitions. The greatest significance that teachers assigned to these competitions was due to the proposed challenges. For teachers, these are real problems that deviate from those that usually appear in textbooks.

They find them suitable for their students; they feel the willingness to solve and discuss the problems with their colleagues and see them as a useful resource for their teaching practice. The teachers that we interviewed considered the problems appropriate to the diversity of students in the classroom, which is an idea that corroborates our claims that the proposed challenges are moderate, and these competitions are inclusive. Note that this view of the teachers is in line with the diversity of students participating in the competitions in terms of school performance, as confirmed by the results of the questionnaire that we have reported. This is an added value of these competitions compared to the most exclusive ones intended for the gifted students.

We know however that the involvement of teachers varied from case to case and from school to school. We have identified and interviewed teachers with high involvement, who have accompanied their students from the Qualifying phase to the Final that revealed their ways of working on the problems either in mathematics classes or in other moments, such as supervised study classes. Noteworthy is the way in which they provide support to the students, by permanently encouraging them to arrive themselves to the solution.

The different degrees of involvement that we identified in teachers do not diminish the role of the teachers as fundamental to the dissemination of these competitions among their students (Fig. 8.10). The vast majority of the mathematics teachers collaborated in the announcement of the competitions by distributing flyers and giving information in their classes. Others went further and gave timely assistance when students requested it. Finally, there are those who actively engaged themselves and closely followed the competitions and took advantage of them for the mathematics learning of their pupils; many of the schools in the region saw the

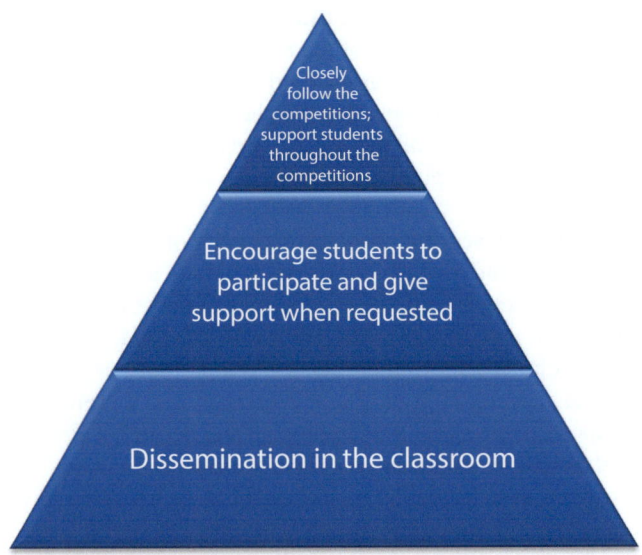

Fig. 8.10 Teachers' levels of involvement in the competitions

SUB12 and SUB14 competitions as curriculum enrichment projects and attested to that in their annual activity plans.

Even in cases where teachers did not introduce technologies in their mathematics teaching process but worked on the problems of the competition in the classroom using other means, there was a positive and open attitude towards the use of technology by the students in their problem-solving and expressing for the competitions. Moreover, there were teachers who used the problems to teach mathematics to their students by resorting to technological tools. In short, the involvement of teachers in the competitions was a lever for a closer relationship between problem-solving and the use of technology.

8.8 Implications and Suggestions for Further Research

The research carried out during our Problem@Web project clearly shows that today's youngsters are able to put together into action their technological and mathematical skills when they have the opportunity and the freedom to decide on the digital tools to use and how to make them work productively. Yet our project shows above all that we are not that well-informed about how this happens. We know that the most appropriate way to promote the development of technological and mathematical skills is to work towards an articulated and comprehensive knowledge on the use of digital tools for mathematical purposes. However, the use of technology for the learning of mathematics in today's schools still means the overcoming of many barriers.

What we have seen in our results is that youngsters do use several digital tools they have access to for addressing and expressing their solutions to non-routine mathematics problems. Moreover, when we look closely at the way they do this, we realise that they are able to develop this articulated knowledge and discover the suitability of the tools to their ways of thinking mathematically. We have put the emphasis on these different and relevant ways of thinking by focusing on the analysis of their problem-driven conceptual models. From this point, we may need to start thinking about technology problem-driven conceptual models to the extent that technology is an integral part of the approach to a problem and that the technology representational expressivity increases the possibilities to think, communicate and represent in mathematics.

In the digital era, we cannot conceive problem-solving as indifferent to the use of the technological tools we have at our disposal. Today, in any profession and in any activity, we use technology to solve the problems of everyday life. The learning of mathematics and, in particular, problem-solving cannot turn off the digital tools. The results we have obtained reinforce and corroborate the numerous recommendations for the use of technology in mathematics learning. Yet our results also give some clues as to what we can do. We can start by thinking about broadening the notion of problem-solving and its role and place in the learning of mathematics.

It is worth mentioning the challenge that teachers face in trying to provide real and stimulating problems that are accessible to the diversity of the students in the classroom. The problems from the SUB12 and SUB14 competitions were seen by the teachers we questioned as distinct from the content-reinforcement problems such as those more commonly presented in textbooks. Indeed, these teachers recognised the need for this more creative, non-routine, open to connections and challenging form of problem-solving. Our findings reinforce the need to truly implement the kind of problem-solving in school that requires individuals to be able to mobilise their mathematical knowledge and technological knowledge, not as separate or compartmentalised but as united. Support for this sort of development can be found in the findings of a recent business survey by the Economist Intelligence Unit (2015), which had problem-solving as the top skill that companies said that their employees needed, with this need expected to grow in importance over the coming years, and also in the recent announcement from the OECD's Programme for International Student Assessment (PISA) that problem-solving is now a PISA assessment domain (OECD, 2013).

It seems important that the school system acknowledges the increasing number of available activities taking place beyond the school as partnership opportunities and a means of curriculum enrichment activities for students. Teachers have shown they are willing to, and manage to, turn those proposals into good resources for their mathematics classes, particularly by turning ideas into challenging and suitable tasks for their students. Therefore, it is important to think of ways of improving and encouraging collaboration with teachers by consulting and involving them. Activities such as the SUB12 and SUB14 mathematical problem-solving competitions are cherished and supported because of the contribution they offer to schools and to parents and their youngsters. At the same time, research should look into these beyond-school resources and lead the way in helping teachers develop strategies to make bridges between beyond-school environments and the classroom. This implies, for example, a rethinking of the ways to disseminate and recommend problem-solving with the use of technology to teachers and schools.

Another aspect that we should ponder is related to the need to create places or periods in schools when students can freely try solving problems with technology without the limitations of time. Problem-solving is the kind of activity that requires experimentation, exploration, investigation, trial, reflection and discussion, something not always fully considered in the school curriculum. School priorities can all too often be overly concerned with time-limited tasks, with the consequence that there is limited possibility to do things differently and to be creative and innovative. The formation and nurturing of mathematics clubs or other additional activities (as in the case of supervised study lessons) where students have the opportunity to participate in projects such as inclusive mathematical competitions of a similar nature to SUB12 and SUB14 are thus possible recommendations.

Notwithstanding the strong and consistent research field on mathematical problem-solving built during the latter part of twentieth century, which helped us dive into this new context of problem-solving, the research on problem-solving is today facing new challenges with the use of digital tools. We can say that the new digital context considerably changes the problem-solving process as we have highlighted in this book. In connection to mathematical thinking with technology, there are new representational forms supported by technological tools, and there is a new digital-mathematical discourse and a mathematical creativity that must be studied.

The conceptualisation of new frameworks to describe and characterise problem-solving with technology seems to deserve greater attention from researchers. A particular question is whether "understand, plan, implement and verify" fit well to the processes that young people perform when they solve problems with the digital tools that they pick.

The research carried out in the Problem@Web project has succeeded in observing and examining the use of digital tools in the classroom for solving some of the problems of the competitions. It even managed to enter the home environment of some of the participants and their families. Yet the study of this complex activity—solving and expressing the solution of a non-routine problem—needs new empirical contexts. Some of the ways to continue forward is to think of conducting focused observations of students in a laboratory environment or widening the scope to virtual interactions between researchers, teachers and the students while they solve problems with digital tools. Beyond that, we are quite aware that from the theoretical and analytical points of view, there is a large horizon to explore in searching for ways to integrate the social and emotional aspects involved in problem-solving (Carreira, Amado, Jones, & Jacinto, 2014). This is a huge challenge that further research may also consider.

The milieu of the SUB12 and SUB14 online problem-solving competitions is a very rich and broad one, in its multiple facets and variety of undergoing motivations, relationships, social groups and perspectives. Searching and inquiring the views of parents and other groups such as policy-makers or educational sponsors will certainly be important. It is also essential to know more about the contribution of mathematical competitions and other similar undertakings to the learning of mathematics. Some evidence of the contribution to mathematics learning from beyond-school activities is provided by Barbeau and Taylor (2009) and more recently by the project reported by Sullenger and Turner (2015). A key issue for further research is how these mathematical activities beyond the school, which welcome and favour technology usage, can be harnessed and help to promote the success of youngsters in mathematics in our digital era.

References

Anderson, J., Chiu, M.-H., & Yore, L. (2010). First cycle of PISA (2000–2006): International perspectives on successes and challenges, research and policy directions. *International Journal of Science and Mathematics Education, 8*, 373–388.

Barbeau, E. J., & Taylor, P. J. (Eds.). (2009). *Challenging mathematics in and beyond the classroom: An ICMI Study*. Berlin: Springer.

Barrera-Mora, F., & Reyes-Rodríguez, A. (2013). Cognitive processes developed by students when solving mathematical problems within technological environments. *The Mathematics Enthusiast, 1–2*, 109–136.

Biagi, F., & Loi, M. (2013). Measuring ICT use and learning outcomes: Evidence from recent econometric studies. *European Journal of Education, 48*, 28–42. http://dx.doi.org/10.1111/ejed.12016.

Borba, M., & Villarreal, M. (2005). *Humans-with-media and the reorganization of mathematical thinking: Information and communication technologies, modeling, experimentation and visualization*. Berlin: Springer.

Carlson, M. (2002). Physical enactment: A powerful representational tool for understanding the nature of covarying relationships? In F. Hitt (Ed.), *Representations and mathematics visualization* (pp. 63–77). Special Issue of PME-NA & Cinvestav-IPN.

Carlson, M., Jacobs, S., Coe, E., Larsen, S., & Hsu, E. (2002). Applying covariational reasoning while modeling dynamic events. *Journal for Research in Mathematics Education, 33*(5), 352–378.

Carlson, M., Larsen, S., & Lesh, R. (2003). Integrating a models and modeling perspective with existing research and practice. In R. Lesh & H. Doerr (Eds.), *Beyond constructivism: Models and modeling perspectives on mathematics problem solving, learning, and teaching* (pp. 465–478). Mahwah, NJ: Erlbaum Associates.

Carreira, S., Amado, N., Jones, K, & Jacinto, H. (Eds.) (2014). *Proceedings of the Problem@Web International Conference: Technology, Creativity and Affect in Mathematical Problem Solving*. Faro, Portugal: Universidade do Algarve. http://hdl.handle.net/10400.1/3750.

Economist Intelligence Unit. (2015). *Driving the skills agenda: Preparing students for the future*. London: The Economist. http://www.economistinsights.com/sites/default/files/Drivingtheskillsagenda_0.pdf.

English, L. (2008). Interdisciplinary problem solving: A focus on engineering experiences. In M. Goos, R. Brown, & K. Makar (Eds.), *Navigating Currents and Charting Directions (Proceedings of the 31st Annual Conference of the Mathematics Education Research Group of Australasia)* (pp. 187–193). Brisbane: MERGA.

English, L., & Sriraman, B. (2010). Problem solving for the 21st century. In B. Sriraman & L. English (Eds.), *Theories of mathematics education: Seeking new frontiers. (Advances in mathematics education)* (pp. 263–290). Heidelberg: Springer.

Fereday, J., & Muir-Cochrane, E. (2006). Demonstrating rigor using thematic analysis: A hybrid approach of inductive and deductive coding and theme development. *International Journal of Qualitative Methods, 5*(1), 80–92.

Fraillon, J., Ainley, J., Schulz, W., Friedman, T., & Gebhardt, E. (2014). *Preparing for life in a digital age: The IEA International Computer and Information Literacy Study International Report*. Heidelberg: Springer. doi:10.1007/978-3-319-14222-7.

Gravemeijer, K. (2002). Preamble: From models to modeling. In K. Gravemeijer, R. Lehrer, B. Van Oers, & L. Verschaffel (Eds.), *Symbolizing, modeling and tool use in mathematics education* (pp. 7–22). Dordrecht, The Netherlands: Kluwer.

Gravemeijer, K. (2007). Emergent modelling as a precursor to mathematical modelling. In W. Blum, P. L. Galbraith, H.-W. Henn, & M. Niss (Eds.), *Modelling and applications in mathematics education. The 14th ICMI Study* (pp. 137–144). New York, NY: Springer.

Hegedus, S., Donald, S., & Moreno-Armella, L. (2007). Technology that mediates and participates in mathematical cognition. In D. Pitta-Panzi & G. Philippou (Eds.), *Proceedings of CERME 5* (pp. 1419–1428). Cyprus: University of Cyprus.

Hegedus, S., & Moreno-Armella, L. (2009a). Intersecting representation and communication infrastructures. *ZDM: International Journal on Mathematics Education, 41*, 399–412.

Hegedus, S., & Moreno-Armella, L. (2009b). Introduction: The transformative nature of "dynamic" educational technology. *ZDM: International Journal on Mathematics Education, 41*, 397–398.

Kadijevich, D. M. (2015). A dataset from TIMSS to examine the relationship between computer use and mathematics achievement. *British Journal of Educational Technology*. http://dx.doi.org/10.1111/bjet.12309.

Lesh, R. (1981). Applied mathematical problem solving. *Educational Studies in Mathematics, 12*, 235–364.

Lesh, R. (2000). Beyond constructivism: Identifying mathematical abilities that are most needed for success beyond school in an age of information. *Mathematics Education Research Journal, 12*(3), 177–195.

Lesh, R., & Doerr, H. (2003a). Foundations of a models and modeling perspective on mathematics teaching, learning, and problem solving. In R. Lesh & H. Doerr (Eds.), *Beyond constructivism – models and modeling perspectives on mathematics problem solving, learning, and teaching* (pp. 3–33). Mahwah, NJ: Erlbaum Associates.

Lesh, R., & Doerr, H. (Eds.). (2003b). *Beyond constructivism: Models and modeling perspectives on mathematics problem solving, learning, and teaching*. Mahwah, NJ: Erlbaum Associates.

Lesh, R., & Zawojewski, J. (2007). Problem solving and modeling. In F. Lester (Ed.), *Second handbook of research on mathematics teaching and learning* (pp. 763–804). Charlotte, NC: Information Age Publishing.

Moreno-Armella, L., & Hegedus, S. (2009). Co-action with digital technologies. *ZDM: International Journal on Mathematics Education, 41*, 505–519.

Moreno-Armella, L., Hegedus, S., & Kaput, J. (2008). From static to dynamic mathematics: Historical and representational perspectives. *Educational Studies in Mathematics, 68*, 99–111.

Moreno-Armella, L., & Sriraman, B. (2010). Symbols and mediation in mathematics education. In B. Sriraman & L. English (Eds.), *Theories of mathematics education: Seeking new frontiers* (pp. 213–232). New York, NY: Springer.

OECD. (2013). *PISA 2012 assessment and analytical framework: Mathematics, reading, science, problem solving and financial literacy*. Paris: OECD Publishing. http://dx.doi.org/10.1787/9789264190511-en.

Santos-Trigo, M. (2004). The role of technology in students' conceptual constructions in a sample case of problem solving. *Focus on Learning Problems in Mathematics, 26*(2), 1–17.

Santos-Trigo, M. (2007). Mathematical problem solving: An evolving research and practice domain. *ZDM: International Journal on Mathematics Education, 39*, 523–536.

Sullenger, K. S., & Turner, R. S. (Eds.). (2015). *New ground: Pushing the boundaries of studying informal learning in science, mathematics, and technology*. Rotterdam: Sense Publishers.

Törner, G., Schoenfeld, A. H., & Reiss, K. M. (2007). Problem solving around the world: Summing up the state of the art. *ZDM, 39*(5–6), 353.

Zeytun, A., Çetinkaya, B., & Erbaş, A. (2010). Mathematics teachers' covariational reasoning levels and predictions about students' covariational reasoning. *Educational Sciences: Theory and Practice, 10*(3), 1601–1612.

Afterword

In 2013 I had the opportunity of reading the colourful and vivid booklet entitled *Um olhar sobre uma competição matemática na web: Os SUBs* (A look over a mathematical competition in the web: The SUBs), written by the organising team of the Portuguese regional competitions called SUB12 and SUB14 (Carreira et al., 2012). This was my first contact with these particular competitions. Written in a friendly and deeply sensitive way, this publication showed how the environment of these competitions could be considered as a rich landscape for researching about mathematical problem-solving, taking into account many questions, perspectives and actors. In this context, the Problem@Web research project emerged and it was developed alongside the competitions.

In 2014 I had a second opportunity to know more details about the Problem@Web project when I attended the Problem@Web International Conference held at Vilamoura between 2nd and 4th of May (Carreira, Amado, Jones, & Jacinto, 2014). On this occasion three strands were proposed as main themes for the conference; technology, creativity and affect in mathematical problem-solving. Rich data extracted from the SUB12 and SUB14 competitions concerning each one of these strands were presented during the conference, among them are amazing solutions and beautiful explanations of the youngsters, emotive words from their parents and reflexive involvement of the students' mathematics teachers.

Finally, I had a third chance to know more about the Problem@Web project, namely, an invitation to write the afterword for this book.

The authors declare that their purpose is "…to provide a contribution to understanding the future of education through analysing the way that the digital generation tackles mathematical problems with the technologies of their choice at the time of their choice" (p. 2). In particular, they analyse and discuss the ways the youngsters use different digital technologies to solve and to express the solutions of some mathematical problems. The examples included in the book address the stated purpose, allowing the reader to imagine school scenarios in which collectives of students with technologies solve problems and express their thinking.

The authors declare a theoretical stance that coherently intertwines different resonant perspectives related to (a) problem-solving as "a concurrent process of mathematization and of expressing mathematical thinking" (p. 107), (b) multiple external representations as ways of supporting learning and providing strategies for solving problems and (c) the symbiotic relation between humans and digital tools, considering the notions of humans with media and co-action with digital tools. Theoretical considerations about the role of representations and technologies during problem-solving and expressing activities provide a solid framework to analyse students' solutions and thinking.

The authors selected different mathematical problems that address three major concepts identified as invariance, quantity variation and co-variation, respectively, emphasising geometrical thinking, algebraic thinking and co-variational reasoning. The students' solutions for these problems were analysed using the theoretical framework, and this allowed the authors to characterise the conceptual models developed by the students, to see the ways in which the uses of digital technologies were related to those models, to know how the affordances of a technological tool influence the problem-solving process and to see the representations designed and the strategies produced by collectives of students with media while solving and expressing mathematical problems.

During the reading of these pages, I paused and spent time solving some of the problems presented in the book, I also discovered new ways of thinking about other problems and once more I verified the amazing things youngsters are able to do if we let them make their own mathematical decisions and choose different digital media to solve and express their solutions.

As I said above, I had three opportunities of approaching the SUB12 and SUB14 competitions and the Problem@Web project. Based on my different experiences during these three opportunities, I emphasise some special features of the competitions and summarise some aspects of the research project in order to raise some special keywords that, in my opinion, represent the spirit of the competitions and the project, aside of the obvious keywords: *problem-solving* and *digital technologies*. Such singular keywords are *inclusiveness*, *supportiveness*, *mathematical communication*, *freedom* and *inspiration*, and I think they should be seriously considered in any mathematics learning environment.

- The SUB12 and SUB14 competitions started in 2005 and invited fifth to eighth grade students (10–14-year-old students) from the southern regions of Portugal to participate in solving a set of mathematical word problems and sending their solutions through the Internet. During the first online phase, the students solved one problem every 15 days, over a period of 6 months. At the end of this first phase, there was a final on-site individual competitive phase that takes place at the Universidade do Algarve (Portugal). These competitions have special features that, from my point of view, make them distinctive from other mathematical competitions, and I would like to highlight some of them: (1) they are online competitions, which means that students from different places can participate without attending a predetermined institution to get and solve the problems; (2) the students can work in groups, which means that the competitions admit collaboration among pairs; (3) the students have enough time to solve the problems, which means that they are not pressed to give quick answers; (4) the students receive friendly feedback, suggestions or recommendations from the organisers if the answer is not correct, which means that the students can rethink their answers and learn from their mistakes; and (5) the students can freely select the digital media to solve the problems and present their solutions, which means that the students have the opportunity of using the technological devices at their disposal. All these features are evidence of a deep concern with *inclusiveness* and *supportiveness* in mathematics education and show the educational positions of the teachers and researchers involved in the organising team of the competitions. These are the first two keywords that I associate with the SUB12 and SUB14 competitions.

- The online environment of the SUBs competitions implies that the students had to send their answers through the Internet, and this fact opens a scenario that invites new digital media to help the youngsters, not only to send the solutions but also to solve, express and write their solutions. The students were always told: *Do not forget to explain your problem-solving process*. This request allowed the organisers of the competitions to understand the students' solutions. At the same time, it allowed students to reflect about their solutions and make the effort of writing an explanation of their procedures and strategies to someone who was not directly observing their solving process. From my point of view, this is an excellent exercise of mathematical communication which is not always common in a mathematics classroom and which is a natural request in this online environment. In this way, the development of written mathematical communication skills was an essential feature of the competitions, and so for me, *mathematical communication* is another keyword in the context of the competitions. All the students' written explanations of their problem-solving processes are a very rich database for the researchers and teachers and allow them to become aware of the youngsters' mathematical potentialities.
- The youngsters participating in SUB12 and SUB14 were instigated and challenged by the problems, and they tried to solve them in many creative ways using technologies. From my perspective, the results presented in the book give clues about the necessity of creating a learning environment inside the mathematics classrooms to include all students in problem-solving activities in collaboration with their colleagues, having enough time to think and choosing the technologies they would like to use. What this book shows is that the students will always surprise us if we give them freedom to make decisions. The more freedom they have, the more creative they become. So, for me, *freedom* is another keyword that I associate with the environment of the competitions.
- The teachers interviewed in this study recognised that the problems posed in the competitions were suitable for all students and useful as pedagogical resources. They also considered that the problems were challenging and different from the well-known applied problems proposed in the school, in which the students know in advance the mathematical topic to be used in order to solve them. So, I think that this book is also an inspiring source for teachers to create new learning scenarios in their classes so that students can work with the special type of freedom enjoyed by the competition participants. In this way, I think that *inspiration* is another keyword that I associate with the Problem@Web project.

I am sure that this book can inspire mathematics teachers and researchers to think about ways of creating inclusive, supportive and free learning environments in school contexts, in which the students can develop and promote mathematical communication skills while solving problems and expressing their thinking with different digital technologies.

To finish this afterword, I would like to make some brief considerations related to possible future research issues.

The delightful and amazing students' mathematical productions analysed in the book made me think about new research questions in the context of the SUB12 and SUB14 competitions or in any context in which students solve problems and also have access to digital technologies. Why do the students decide to choose certain software to solve a given problem? What are the characteristics of those problems that the students prefer to solve using digital tools?

The book also offers evidence about how the use of digital tools allows youngsters to solve problems before having the specific mathematical knowledge that the teachers think they need to solve such problems. This characteristic of digital environments offers the possibility of thinking about a curricular reorganisation at schools. New problems, such as those coming from SUB12 and SUB14, could be proposed at schools if digital technologies were understood as a tool to mathematically think with. Which are the necessary conditions at schools to incorporate, in the daily school activities, problems such as those proposed at the SUBs competitions?

Regarding the organisation of the competitions, I wonder: what would happen if the students were allowed to use digital technologies at the final phase of the competitions?

Finally, I propose to look the other way around and ask: which problems would the youngsters pose if they knew they could use digital resources to solve them? Problem posing within digital environments inside or outside the school context seems to be a challenging landscape to research.

Reading the book and writing this afterword was a pleasant experience for me, considering *experience* in the sense of the Spanish author Jorge Larrosa (2003): "a journey in which the starting point was ordinary, familiar, and well-known for me, but when I left it, I arrived to a strange and unknown place that surprised me and from which I came back, transformed, to my original place".

Mónica E. Villarreal

Córdoba, June 2015

References

Carreira, S., Amado, N., Ferreira, R. A., Silva, J. C., Rodriguez, J., Jacinto, H., et al. (2012). *Um olhar sobre uma competição matemática na Web: Os SUBs*. Faro, Portugal: Universidade do Algarve. http://hdl.handle.net/10400.1/2733.

Carreira, S., Amado, N., Jones, K., & Jacinto, H. (Eds.) (2014). *Proceedings of the Problem@Web International Conference: Technology, Creativity and Affect in Mathematical Problem Solving*. Faro, Portugal: Universidade do Algarve. http://hdl.handle.net/10400.1/3750.

Larrosa, J. (2003). *La experiencia de la lectura*. México: Fondo de Cultura Económica.

About the Authors

Susana Carreira is an Associate Professor at the Faculty of Sciences and Technology of the University of Algarve, Portugal, where she teaches mathematics and mathematics education to graduate and post-graduate students, and an Invited Associate Professor at the Institute of Education of the University of Lisbon, Portugal, where she collaborates in the supervision of master and doctoral students in mathematics education; she is also a member of the Institute of Education Research Unit. At the University of Algarve, she has taken part in the organising team of the annual editions of the web-based regional mathematics competition SUB12® and SUB14®. Her research activity has been on technologies in mathematics teaching and learning, mathematical modelling and applications, problem-solving and mathematical creativity. She has participated in several national funded research projects and was the principal investigator of the Problem@Web Project (2011–2014). Internationally she has developed projects in partnership with Brazil, where she has also co-supervised master students in mathematics education, and more recently she is a member of the EU-funded project MILAGE. She has been actively involved in several national and international conferences, including the *International Conference on Technology in Mathematics Teaching (ICTMT)*, the *International Conference on Teaching Modelling and Applications (ICTMA)* and the *Congress of European Research in Mathematics Education (CERME)*. In the latter she has served as co-leader and as leader of the Thematic Working Group on Applications and Modelling. She is a member of the ICTMA Executive Committee for 2015–2017. She has published research articles in mathematics education and in national and international journals and edited books, particularly related to mathematical modelling, problem-solving, digital technologies in mathematics learning and mathematics competitions.

Keith Jones is Associate Professor in Mathematics Education at the University of Southampton, UK, where he is the Deputy Head of the Mathematics and Science Education (MaSE) Research Centre. His expertise in mathematics education spans geometrical problem-solving and reasoning, the use of technology in mathematics education and mathematics teacher education and professional development. He is on the editorial board of several journals, including *Educational Studies in Mathematics*, the *Journal of Mathematics Teacher Education* and *Research in Mathematics Education*. He co-edited a special triple issue (a complete volume) of *Educational Studies in Mathematics* on proof and proving while using dynamic geometry software, four special issues (a complete volume) of the *International Journal of Technology in Mathematics Education* on research on using technology in mathematics education and a special issue of *ZDM, the International Journal on Mathematics Education*, on research on mathematics textbooks. He has been an active member of the thematic group on *Tools and Technologies in Mathematical Didactics* of the *European Society for Research in Mathematics Education* (ERME) from its inception in 1998; from 2000 to 2003 he co-led the group. He has taken part in several ICMI studies, including ICMI Study 9 on geometry education, ICMI Study 11 on the teaching and learning of mathematics at University level, ICMI Study 17 on digital technologies in mathematics education, ICMI Study 19 on proof and proving in mathematics education and ICMI study 22 on task design in mathematics education. He has served on the international programme committee for several of the ICTMT (*International Conference on Technology in Mathematics Teaching*) conferences. He has led and worked on numerous projects, including participating in several EU-funded projects that have focused on the use of technology in the teaching and learning of mathematics. He has well-established research collaborations with educators in various parts of Europe and in China and Japan. He has published widely, with his recent co-authored book for Oxford University Press being entitled *Key Ideas in Teaching Mathematics*.

Nélia Amado is an Assistant Professor at the Faculty of Sciences and Technology of the University of Algarve, Portugal, and is a member of the Research Unit of the Institute of Education of the University of Lisbon. Over the years she has been involved in the initial and continuing education of mathematics teachers where she has devoted particular attention to the training of future and in-service teachers on the integration of digital technologies in mathematics teaching. She has served as a member of the Advisory Committee appointed by the Portuguese Ministry of Education for the implementation of the national Plan of Mathematics and the renovation of the mathematics curriculum of basic education. More recently she has been assigned the position of external expert of schools that require priority intervention in the improvement of mathematics learning outcomes. Her research and the supervision of master and doctoral students cover the areas of teacher education, the integration of technology in mathematics teaching, the assessment of learning, the affective realm of problem-solving and mathematical competitions. She has participated in national and international research projects, including the Problem@Web project, the EU-funded MILAGE project and bilateral projects with Brazil. She has been involved in the organising and scientific committees of international conferences, particularly the recent *10th International Conference on Technology in Mathematics Teaching* and the *Problem@Web International*

Conference. She has published articles in national and international peer-reviewed journals and is a co-editor of a special issue of the journal *Teaching Mathematics and its Applications.*

Hélia Jacinto is an elementary and secondary mathematics teacher at José Saramago Group of Schools, Palmela, Portugal. She received the M.Sc. degree in Educational Technology from the Portuguese Catholic University and is currently pursuing the doctoral degree in Mathematics Education at the Institute of Education of the University of Lisbon, where she is also developing her research activities as a member of the Research Unit of the Institute of Education. She is an active member of the Portuguese Association of Mathematics Teachers and of the coordinating committee of its Research Working Group. She has been engaged in mathematics teacher training for several years, namely, in the national Plan of Mathematics where she also offered support to regional primary and elementary schools in the implementation of a new mathematics syllabus. Over the past years, she has organised several national conferences, serving either in their scientific or organising committees, and she has also been a member of the organising committee of two international conferences held in Portugal, namely, the *Problem@Web International Conference* and the *10th International Conference on Technology in Mathematics Teaching*. Her research interests include the development of conceptual models in mathematical problem-solving and the use of digital technologies in the learning of mathematics, with particular focus on problem-solving, covering both school and beyond-school settings of which she has been researching mathematical problem-solving competitions under the scope of the Problem@Web research project.

Sandra Nobre is an elementary and secondary mathematics teacher at Prof. Paula Nogueira Group of Schools, Olhão, Portugal. She received her M.Sc. degree in Mathematics for Teaching from the University of Algarve. She was awarded a scholarship by the Foundation for Science and Technology to develop her Ph.D. dissertation in Mathematics Education at the Institute of Education of the University of Lisbon. At this university she is also developing her research activities and is a member of the Research Unit of the Institute of Education. Her research mainly focuses on the learning of algebra and algebraic thinking; she is particularly interested in studying the ways of representing and expressing algebraic thinking of students from middle school through tasks that involve solving problems and the use of the spreadsheet. Over the years she has been involved in teacher training, namely, in teacher professional development programmes in primary education concerning the implementation of curricular reforms in mathematics. She has also participated in the national Plan of Mathematics, where she has worked with groups of mathematics teachers in the preparation and carrying out of didactical and school-based projects in several schools. She has regularly collaborated with the organising team of the SUB12 and SUB14 mathematics competitions, and she has participated as a member of the research team of the Problem@Web project. Her contribution to the project covered mainly the study of students' problem-solving skills and strategies and their ways of representing and expressing mathematical thinking in solutions based on the use of digital technologies.

Index

A
Algebraic representations, 183, 184, 186, 188
Algebraic symbolism, 144, 145, 149
Algebraic thinking
  algebraic structures, 144
  and arithmetic field, 144
  and symbolism, 144
  context and method, 147–149
  data analysis (*see* Data analysis, algebraic thinking)
  design principles, 142
  development, 145–146
  generational activities, 144
  global/metalevel activities, 144
  spreadsheet (*see* Spreadsheet)
  technological tools, 144
  transformational activities, 144
Algebraic type of representation, 192–194
Algebraic/symbolic representation, 186, 189
Arithmetic representations, 178

C
Camtasia Studio, 148
Challenging Mathematics In and Beyond the Classroom, 62
Co-action, 98
  conceptual models, 229
  digital tool, 225
  humans-with-media (*see* Humans-with-media)
  with spreadsheet
    computational development, 146
    continuous dynamics, 146
    digital technology, 146
    GeoGebra, 146
    intermediate relations, 146
    numerical data, 147
    problem-solving, 147
    student and tool, 147
Common usage digital tools, 204
Computer protocol, 148
Conceptual models, 113, 242
  activities, 227
  affordances, digital tools
    cell B19, 116
    ecological, 114
    GeoGebra, 116
    graphics view, 115
    interaction, 115
    invariant, 114
    mathematics education, 117
    spreadsheet view, 116
    SUB12 and SUB14, 115
    SUB14, 115
    technological tool, 114
  completed journey, 183
  context and method
    GeoGebra, 123
    geometrical problem, 123
    problem-solving process, 123
    SUB14, 123, 124
    wider frameworks, 122
  co-variation, 199
  developing journey, 183
  developing students', 175
  DGS, 120–122
  humans-with-media, 226
  humans-with-media mathematising
    concepts and structures, 120

Conceptual models (*cont.*)
   expressions, 120
   human activity, 119
   mathematical concepts, 119
   mathematical techniques, 119
   problem-solving process, 120
  mathematical knowledge, 227
  and mathematical thinking, 174
  participants, 226
  participants' problem solving and expressing
   codification of students' solutions, 191
   completed journey, 183
   developing journey, 188
   pictorial/figurative representations, 189
   quasi-algebraic language, 188
   tabular representations, 189
   textual/descriptive representations, 188, 189
  problem-driven, 204
  problem-solving, 85
  productive thinking, 88
  solutions, 226
  SUB14, 227
  subject and context
   graphic calculator, 118
   human mind, 117, 118
   humans-with-dynamic-geometry, 118
   humans-with-media, 117, 118
   ideas influence, 118
   mathematical thinking, 117, 118
   spreadsheet, 118
  theoretical ideas
   analysis and critical skills, 113
   digital technologies, 113
   mathematical knowledge, 113
   technological tools, 114
  and types of mathematical representations, 180
  young participants, 84
Continuous dynamics, 146
Co-variation, 230, 231
Co-variational reasoning
  carriers of metaphorical meaning, 176
  complexity, 175
  composite functions, 177
  coordination of variables changing, 175
  distance travelled and displacement, 176
  dynamic functional situations, 175
  mathematical reasoning, 175
  in mathematics teaching and learning, 175
  mental imagery, 176
  real and kinaesthetic images, 176
  variables and functions, 176
  verbal description, 177
  visual forms, 177
  visual registers, 177
  visualisation component, 178
  written description, 177
Creative representations, 180, 206

D
Data analysis
  building flowerbed, 124–125
  comparison and contrast
   activities, 138
   digital solutions, 138
   fixed dimension, 136
   GeoGebra, 135
   humans-with-media, 135
   images and ideas, 135
   invariance, 137, 138
   knowledge and procedures, 135
   measurement tools, 136–137
   problem-solving activity, 137, 138
   solution process, 136
   students-with-media, 137
   visible entities, 136
  conceptual models (*see* Conceptual models)
  definition of categories, 186–187
  experts' solutions to problem
   algebraic algorithmic method, 182–183
   algebraic approach, 182
   completed journey, 183, 184
   conceptual models, 183
   developing journey, 185
   tabular representation, 185
  mapping of students' ways of thinking, 187
  representation modes (*see* Representation modes)
  unit of analysis, 188
  zooming
   exhibit A, 125–127
   exhibit B, 127–129
   exhibit C, 129–132
   exhibit D, 132–135
Data analysis, algebraic thinking
  opening of the restaurant "sombrero style"
   algebraic language, 160
   algebraic language of students' model, 163
   Ana's model, 166, 167
   Ana's solution, 166
   Carolina's model, 165
   Carolina's solution, 164
   Gil's solution, 161
   Maria and Jessica's solution, 162, 163
   SUB14, 160
   translation of Gil's model, 161, 162

Index

treasure of King Edgar
    Abel, Bruno and Carlos's solution, 150
    Ana's solution, 156
    David's solution, 157
    David's translation, 158, 159
    hierarchy of analysed solutions, 159
    Marcelo's solution, 152
    possible algebraic approach, 150
    SUB14 competition, 149
    translation of Ana's model, 157
    translation of group's model, 151
    translation of Marcelo's model, 153, 154
    variable-column, 159
Diagrammatic representation, 202, 203
Digital media, 243
Digital technologies, 1–3, 6, 8, 18, 35–46
    characteristics, 32
    communication, 30
    competitions, 28
    components, 32
    creating tables, 36
    digital tools, 28, 29
    e-mail, problems, 26
    fluency, 26
    geometrical constructions, 29
    geometrical software
        digital tool mirror, 47
        digital tools, 51
        drawing tool, 47
        GeoGebra, 45, 47
        GeoGebra's graphic view, 51
        graphs, 47
        systematic counting, 48
        technological skills, 51
    images and diagrams, 29
        calculations, 42
        mathematical problems, 39
        motion problem, 41
        problem-solving routines, 40
        skills, 40
        slides, 43
        tools, 43
        visual expression, 42
    internet and e-mail, 27
    internet connection, 27
    numbers and formulas, 29
    numerical software
        algebraic problem, 46
        calculations, 45
        formulas, 44, 45
        graphics, 44
        levels, 44
        problem-solving, 45
        spreadsheet, 44

    pie charts, 32
    problems, 30
    software tools, 29
    solutions, 32
    solving process, 30, 32
    SUB12, stages, 34
    symbolic language, 35
    symbols and mathematical operations, 29
    tables, 29
        counting, 37
        menus, 36
        option, 36
        spreadsheet, 36, 38
        SUB12 and SUB14, 35
        text editor, 35
        Word and Excel, 35
    technical skills, 34
    text, 29
    text boxes, 33
    webpage, 26
    website, 26
Digital tools
    activities, 236
    challenges, 236
    mathematics learning, 237
    non-routine mathematics problems, 235
    SUB12 and SUB14 competitions, 236
    technological and mathematical skills, 235
Digital-mathematical discourse, 217, 224, 232, 233, 237
    algorithmic style, 91
    assignments, 90
    communication, 92
    expression, 91
    learning activity, 90
    mathematical ideas, 91
    mathematical thinking, 90
    SUB12 and SUB14, 93
    symbolism, 91
    VMT, 92
Dynamic geometry software (DGS)
    digital technologies, 120
    implicit dynamism, 121
    mathematical thinking, 122
    measurement activities, 122
    mental activity, 121
    solving practical, 121
    theoretical concepts, 120

E

External representations
    characteristics, 96
    mathematical activity, 97

External representations (*cont.*)
  participants, 97
  spatial tasks, 98
  strategies, 96
  SUB12 and SUB14, 96
  tasks, 96

F
Figurative/pictorial representations, 206

G
GeoGebra
  construction, 127, 132, 136
  use, 45–51, 115, 127
Geometrical invariance, 113, 127, 129, 135–138
Geometry problem, 47–51, 113, 121–123, 216, 226
Graphical representation, 183

H
Humans and digital tools, 242
Humans-with-media
  algebraic models, 99
  artifact-mediated-activities, 106
  cognitive activity, 99
  conceptual models, 226
  concrete visual and iconic approach, 101
  dichotomous vision, 100
  digital tools, 100, 106
  dynamic digital objects, 102
  dynamical tools, 105
  empirical context, 106
  experimental approaches, 99
  interpretation, 100, 102
  knowledge and cognition, 104
  Leonor's diagram, 100, 101
  mathematical activity, 99
  mathematical procedures, 98
  software tools, 100
  solution, 99
  SUB12, 102, 103, 105
  SUB14, 100
  theoretical perspectives, 225
  transformative nature, 98

I
ICILS. *See* International Computer and Information Literacy Study (ICILS)
IMO. *See* International Mathematical Olympiad (IMO)

Inequality solving techniques, 148
Intermediate relations, 145
International Computer and Information Literacy Study (ICILS), 2, 3
International Mathematical Olympiad (IMO), 65
Inventing representations, 180

M
Mathematical competitions
  activities, 23
  demography, 23
  digital technologies, 25
  computer and internet, 25
  problems, 25
  qualifying phase, 26
  SUB12 and SUB14, 173
Mathematical knowledge, 244
Mathematical problem-solving, 63, 65–72, 241
  activities, 58
  advertising, 58
  Challenging Mathematics In and Beyond the Classroom, 62
  classes, 52, 59, 60
  communication
    characteristics, 74
    essential feature, 72
    knowledge, 74
    requirement, 72
    resolution, 74
    strategies, 73
  community and young participants, 63, 64
  competitions and enrichment activities, 63
  difficulties, 61
  digital fluency, 21
  digital tools, 52
  discussion, 61
  e-mail, 59
  general tools, 21
  information, 58
  learning, 63
  Likert-type questions, 22
  literature, 63
  mathematical works, 52
  merit diploma, 60, 61
  motivation, 62
  participants, 62
  procedures, 62
  receive awards, 62
  sketches, 51
  SUB 12 and SUB14, 21, 22, 24
    activities, 69
    challenges, 63, 66, 71
    characteristics, 68
    classroom, 66, 68, 70

content application problems, 69
deep-rooted patterns, 70
development of technological skills, 65
expectations, 72
feature, 65
guiding principle, 68
learning, 71
literature, 67
strategies, 65
strictly talent-search model, 67
teachers' practices, 70
technological tools, 51
testimonies, 57
webpage, 23
young people, 62
Mathematical problem-solving and expressing
conceptual models, 226, 227
co-variation, 230, 231
digital and asynchronous communication, 214
digital tools, 225
geometry software, 216
interview data, 215
knowledge, 214
mathematical problems, 217
mathematical thinking, 217
numerical software, 216
online communication, 215
participants, 215–217
quantity variation, 227–229
solutions, 215, 216
SUB12 and SUB14, 214
theoretical framework, 223–225
Mathematical representations, 180, 186
Mathematical thinking
digital tools, 85, 96
problem solving, 95
Mathematics teachers' involvement
awarding ceremony, 218, 220
characteristics, 218
classes, 218
communication and expression, 221
competitions, 234
digital technologies, 223
mathematical thinking, 221
participants, 222
participation, 218
pedagogical resource, 220
poster announcing, 218, 219
solution, 220
stages, 223
Mathematization
activities, 86

conceptual model, 86, 88
digital drawing, 86
digital solution, 86
RME, 85
solution, 86, 87
Moderate challenge, 72
Moderate-challenging problems, 210, 211, 217, 218
Motion problem
algebra and algebraic procedures, 205
algebraic methods, 174
and co-variation (*see* Co-variational reasoning)
and visualisation (*see* Visuality)
common usage digital tools, 204
data analysis (*see* Data analysis)
digital solutions, 174
digital tools, 181
digital visuality, 207
displacement and time, 205
documentary data analysis, 180
graphical character, 206
language, 206
launch, 180
modes of explanation, 180
non-formal and non-algorithmic mathematisation, 205
problem solving-and-expressing, young participants, 174
representation (*see* Representation modes)
research purpose, 179
SUB14 (edition 2011–2012), 180

P

Pictorial type of representation, 198
Pictorial/figurative representation, 191, 193, 195, 200
PISA. *See* Programme for International Student Assessment (PISA)
Problem-solving
algebraic, 170
co-action, 146–147
explanation, 156, 164
with spreadsheet, 145–146, 170
Problem solving-and-expressing
activities, 93
capture students' responses, 94
conceptual model, 84, 89
diagrammatic software, 94
digital communication context, 84
digital media, 95
digital tools, 85, 88

Problem solving-and-expressing (*cont.*)
   dynamic geometry software, 90
   essential feature, 95
   mathematical competitions, 84
   mathematical representation, 88
   mathematization, 85, 86, 88
   software functionalities, 90
   theoretical development, 85
   types, 93
   visual expression, 94
Problem@Web project, 21, 241, 242
   aims and methods, 17
   co-action, 213
   data-driven encoding, 213
   digital natives, 1, 2, 18
   digital tools, 2, 16, 18, 232
   focus
      attitudes and emotions, 8
      digital tools, 6
      mathematical thinking and problem-solving, 7
   mathematical problems, 213
      activity, 5
      investigation, 4
      problem-solver, 4
      solving, 4
      SUB12 and SUB14, 5, 6
      tools, 6
   methodological
      analysis, 16
      data sources, 15
      digital tools, 15, 16
      formulas, 16
      frameworks, 17
      GeoGebra/ tables, 16
      mathematical problems, 15
      mathematics classes, 16
      online phases, 16
      online questionnaire, 15
      quantitative and statistical data, 15
      spreadsheet-based solutions, 17
   online problem-solving environment, 18
   participants, 213
   qualitative methodology, 213
   research foci, 212
   routines, 1
   skills, 18
   social communicators, 2
   SUB12 and SUB14, 2
      final phase, 11
      incomplete/faulty answer, 11
      league table, 10
      mathematical challenges, 12
      mathematical reasoning, 12
      non-routine mathematical problems, 11
      qualifying phase, 8, 11
      reformulating and resubmitting answers, 12
      solutions, 11
   technology, young people
      creating/editing documents, 3
      digital natives, 2
      digital resources, 4
      equipments, 3
      software applications, 3
   young participants, 232
Problem-driven conceptual development, 84, 87

Q

Quantity variation, algebraic thinking.
   See Algebraic thinking
Quasi-algebraic language, 188
Quasi-algebraic type of representation, 192, 193
Programme for International Student Assessment (PISA), 236

R

Realistic Mathematics Education (RME), 85, 119
Representation modes, students' digital productions
   algebraic type, 192, 193
   diagrammatic representation, 202–204
   pictorial/figurative representation, 193, 195, 198, 200
   quasi-algebraic type, 192, 193
   quasi-dynamic aspect, 195
   tables/tabular representations, 199, 201
   textual/descriptive representations, 194–196
   textual/descriptive/quasi-algebraic representations, 191–192
   textual/descriptive/quasi-algebraic type of representation, 192
RME. *See* Realistic Mathematics Education (RME)

S

Spreadsheet
   Ana's solution, 166
   and algebraic conditions, 168
   and symbolic language of algebra, 169
   co-action, 146–147
   digital representations

Index

algebra problems, 142
   communicating, 142
   vs. computing environments, 142
   educational environment, 143
   handling, 142
   João's solution, 143, 144
   Monica's solution, 143
   numbers in cells, 142–143
   solving and expressing, 143
   variable-column, 142
   vocabulary, 142
electronic, 141
Gil's solution, 161
into algebraic language
   Ana's model, 157
   Carolina's model, 165
   David's translation, 158, 159
   translation of Ana's model, 167
   translation of Gil's model, 161, 162
into symbolic algebraic language
   translation of Marcelo's model, 154
   translation of the group's model, 151
Maria and Jessica's solution, 163
Monica's solution, 143
natural language, 142
problem-solving, 145–146
Standard scientific representations, 179
Student-invented representations, 180
Students' technological competencies, 218
SUB12 and SUB14, 241–243, . *See also*
   Problem@Web project
Symbolic representations, 186. *See also*

T

Tabular representations, 182, 183, 185, 187,
   189, 191, 199, 201
Teachers' practices, 57, 70
Teachers' views on competitions, 57, 74
   activities, 55
   education cycles, 56
   gaining awareness, 56
   mathematical problem solving
      (*see* Mathematical problem solving)
   participation, 56, 57
   solution, 57
   SUB12 and SUB14, 55

tasks, 56
technology usage (*see* Technology usage)
Technology
   digital, 146
   spreadsheet (*see* Spreadsheet)
   tools, 144, 146, 148
Technology usage
   advantage, 78
   Excel sheet, 78
   National Technological Plan, 76
   numerical/algebraic problems, 77
   participation, 75
   purposes, 75
   solutions, 76
   young participants, 77
Textual representations, 183
Textual/descriptive representations, 188, 189,
   193–197, 206
Textual/descriptive/quasi-algebraic
   representations, 189–191

V

Variable
   algebraic, 158
   dependent, 158, 169
   independent, 143, 153, 155, 158, 169
Variable-column, 142, 156, 159, 166–170
Virtual math teams project (VMT), 92
Visual representations, 177
Visuality
   and coordination, 178
   animation tool, 179
   digital, 207
   digital solutions, 178
   inspection, representational elements, 178
   representational features, 178
   researcher's priority, 179
   Sherin's genetic analysis, 179
   speed and relative positions, moving
      bodies, 178
   visual animations fosters, 179
VMT. *See* Virtual math teams project (VMT)

W

Web-based competitions, 210

MIX
Papier aus verantwortungsvollen Quellen
Paper from responsible sources
FSC® C105338

If you have any concerns about our products,
you can contact us on
ProductSafety@springernature.com

In case Publisher is established outside the EU,
the EU authorized representative is:
**Springer Nature Customer Service Center GmbH
Europaplatz 3, 69115 Heidelberg, Germany**

Printed by Libri Plureos GmbH
in Hamburg, Germany